The Macat Library
世界思想宝库钥匙丛书

解析查尔斯·达尔文

《物种起源》

AN ANALYSIS OF

CHARLES DARWIN'S

ON THE ORIGIN OF SPECIES

Kathleen Bryson　Nadezda Josephine Msindai ◎ 著

贾顺厚 ◎ 译

上海外语教育出版社
外教社 SHANGHAI FOREIGN LANGUAGE EDUCATION PRESS

目　录

CONTENTS

引言

要 点

- 查尔斯·罗伯特·达尔文（1809—1882）是英国的一位博物学家＊（即研究自然世界的学者），也是一位地质学家＊（即研究地球岩石形成和历史的学者）。

- 其著作《论依据自然选择的物种起源》阐释了在"自然选择"＊作用下物种的诞生、演化与灭绝＊（自然选择这一过程指最能适应某一特定环境的生物体存活下来，并且有助于其将生存的特性遗传给下一代，而适应性差的生物体则被淘汰。）

- 《物种起源》中的主要概念对生命科学而言意义重大。

查尔斯·达尔文其人

查尔斯·达尔文是著名的博物学家，他撰写了《论依据自然选择的物种起源》（1859）一书，这是一部用他称为"自然选择"的过程来解释生物进化＊的著作。这部具有非凡影响力的著作颠覆了科学家对地球生命起源与发展演变的认知，作者也因此获得了"进化论之父"的美称。

达尔文于 1809 年出生在英国一个名叫什鲁斯伯里的集镇上。作为镇上声名显赫的达尔文—韦奇伍德家族的一员，达尔文出身富贵，生活优越。1831 年他毕业于剑桥大学的基督学院，获得了学士学位；同年冬，他登上贝格尔号＊考察船，任随船博物学家一职；这是一艘皇家海军舰艇，正要进行一次为期五年的全球科学考察航行。考察结束后，达尔文返回家乡，他的父亲安排了几项以他为受益人的投资计划，由此，达尔文成为一个自筹经费的"绅士科学家"。[1]

达尔文的社交圈子里有专业学者，也有激进的思想家，前者的代表是地质学家查尔斯·莱尔*，后者则有社会理论家哈里特·马蒂诺*。这些人都对达尔文自然选择学说的建立有过帮助。起初，达尔文延期出版自己的新理论，担心会激起公众的愤恨情绪；因为这一理论中关于地球生命的起源问题与基督教的教义背道而驰（这只是一个因素，还有其他原因）。然而1858年，博物学家阿尔弗雷德·拉塞尔·华莱士*写信给达尔文，提出了关于生物进化的相似的理论，这才使达尔文下定决心，将新理论立即发表。

华莱士（1823—1913）是威尔士人，集博物学家、收藏家、热带领域生物学家（热带领域生物学家即热带生物专家）于一身。他与达尔文在1858年共同在伦敦一本专门讨论和研究自然历史的名字叫"林奈学会"*的会刊上发表了一篇依据自然选择的进化理论的文章。次年，达尔文发表了《物种起源》，对该理论作了更加详细的说明。达尔文一生中前后总共撰写出版了25本著作，涉及藤壶、植物及蚯蚓等诸多主题。达尔文将整个生命都献给了科学研究事业；他于1882年4月在肯特郡的达温宅家中逝世，遗体葬于伦敦威斯敏斯特大教堂。[2]

《物种起源》的主要内容

《物种起源》的主要内容是进化过程：地球上所有物种都来源于一个祖先。自然选择学说解释了生物体是如何适应环境，最终生存下来的。这一理论建立在观察之上：虽然各种动植物都倾向于过度繁殖，但赖以生存的资源有限，导致生物数量和规模基本恒定。

为了说明这一点，达尔文指出生物体之间存在着种种细微的差异，尤其是它们固有的行为（他定义为"本能行为"）。由于同一个物种*包含着数量巨大的个体，并不是所有的个体都能够生存下去；为了生存，就有斗争。在争夺有限资源的竞争中，那些最能够适应环境的个体就更容易生存下去，并将其特征（即它们的行为和物理属性的差异）传递下去。某一物种逐渐演化并占据不同的环境，最终形成新的物种，无法与其前代物种进行繁殖。

自然选择这一概念与突变*和遗传漂变*是进化论研究的主要部分。突变是指动物基因组成中一种相对突然的变化，导致其行为或体格的变化；遗传漂变则指大体上，随着时间的推移，动物繁殖时，其遗传信息改变的过程。

《物种起源》中的次要的主题包括性别选择*，物种形成*（即新物种的演化过程），和渐进性*（指物种在漫长的演变过程中的变化方式）。

性别选择（例如通过选择配偶促成的自然选择）可以发生在同性或异性之间。同性选择*指的是同性个体竞争异性配偶。这种情形多见于雄性个体间的竞争。攻击技能强、体格魁梧、或持有最大武器的雄性个体的获胜机会就高。优胜者获得了独占配偶的权利，因此繁衍的后代超过失败者；也就是说，能够决定竞争结果的各种特征如果是遗传的，那就是自然选择发挥了作用。异性选择*，有时候被称为"雌性选择"，指雌性个体基于某些特质选择潜在配偶。如果中意的个体与它的对手在遗传上有所不同，也会产生自然选择。

达尔文认为进化是一个缓慢而渐进的过程。在书中，他引入了渐进这一概念——物种在漫长的过程中不断演变和积累着细微的变化。他还解释了如何通过种群分化的方式产生新的物种；种群一旦分化，就产生了两种新的物种。如果其中一个物种占据大面积的领地，环境发生变化就意味着不同的个体会经受不同的压力，它们的适应性也就有所差别。

在不同地区，因为自然选择在各个环境中独立发挥作用，因此产生新物种。达尔文认为不同区域之间也必定存在混合区域：该区域内的个体因为不能充分适应两个邻近区域中的任何一个生存环境，最终只能走向消亡。

尽管进化过程通常需要很长一段时间，但也有例外；例如，一些病毒和昆虫可能会在相对较短的时间内对某些农药产生抗药性。

《物种起源》的学术价值

《物种起源》是有史以来最有影响力的科学文献之一。书中提出的自然选择理论为复杂的生物进化过程以及这些生物不需要神的干预而精妙地适应环境提供了一种解释。达尔文运用大量广泛的论据，论证了他的变异性遗传 * 理论（就是所谓的进化模式），以及所有生物通过一个共同的祖先而彼此相互关联的观点。

也许更值得一提的是，达尔文撰写的这本书，不论是科学家、还是普通大众，读过之后都可以理解他的理论。然而，读者数量众多，难免引发争议，因为这一著作当时被认为是对国家和教会盛行的意识形态的挑战。

达尔文确立了进化论，奠定了进化生物学 * 领域的学科基础，并且他的理论对许多学科都有一定的影响，比如神学 *（对宗教思想的系统研究，一般通过梳理宗教经文的方式进行）、现代哲学、海洋学 *（对海洋的科学研究）、语言学 *（研究语言结构和性质的学科）、人类学 *（对人类文化和社会生活的研究）、历史学和经济学等。

《物种起源》出版近 160 年后的今天，进化仍然是存在着争议的话题。不少人仍然持怀疑态度，在美国，只有不到一半（48%）的人相信达尔文的自然选择学说。[3] 与此同时，无神论者*（不相信任何神的人）和唯物主义者*（为所有自然现象寻找物质原因的人）认为，该理论对宗教提出了一个严肃而坚决的挑战。进化论和自然选择学说为否定神在创造人类和动物中起了作用这一说法提供了正当理由。

然而，认为这自然世界的奇迹是由神创造的宗教信徒则将这一理论视为一种莫大的侮辱。在谈到实证主义*这一原理，即我们的知识只能通过科学论证的事实得以向前推进的理论时，进化生物学家恩斯特·迈尔*这样写道："把上帝从科学中请出局，就为严格地、科学地解释所有自然现象提供了可能；这就产生了实证主义，引发了一场强大的知识界革命和精神革命，它的影响深远，今天都能感受得到。"[4] 但时至今日，宗教看法依然众说纷纭。有的人接受自然选择的概念，而有的人对达尔文的理论仍持反对的态度。

尽管自《物种起源》出版以来，进化论这一理论有无数的创新发展，但自然选择学说仍然是理解物种变化的一种途径。[5] 达尔文的观点已然影响了科学发展和社会进步。

---

1. 珍妮特·布朗：《查尔斯·达尔文传（第一卷）：航游》，普林斯顿：普林斯顿大学出版社，1996年，第434—435页。
2. 珍妮特·布朗：《查尔斯·达尔文传（第二卷）：地方的力量》，伦敦：皮米里科出版社，2003年，第496页。

3. 皮尤论坛："宗教团体：进化论观点"，2009 年 2 月 4 日，登录日期 2016 年 2 月 5 日，http://www.pewforum.org/2009/02/04/religious-differences-on-the-question-of-evolution/。

4. 恩斯特·迈尔："达尔文对现代思想界的影响"，《美国哲学学会会刊》第 139 卷，1995 年 12 月第 4 期，第 317—325 页。

5. H. 艾伦·奥尔："用遗传学检测自然选择"，《科学美国人》第 300 卷，2009 年第 1 期，第 44 页。

第一部分：学术渊源

1 作者生平与历史背景

要点 🔑

- 进化理论和自然选择机制共同为所有当代生物科学提供了学术框架。

- 查尔斯·达尔文的社会地位，以及他能够经常接近当时一流的知识分子，最终促成了他的理论发展。

- 19世纪中期，欧洲爆发了关于奴隶制、殖民主义和人类天性的后启蒙运动*改革思潮。（启蒙运动强调理性和自由的概念；殖民主义是一个国家或民族通过占领别国的土地，进而对别国进行剥削的过程。）

为何要读这部著作？

查尔斯·达尔文的《论依据自然选择的物种起源》（1859）解释了物种是如何通过自然选择过程从一个共同的祖先演化而来的。在知识进步深受宗教信条限制的年代，达尔文借鉴了各种不同的资料，最终形成了对科学产生重大变革的理论。之后，进化论就成为合法的科学探究领域。[1]

达尔文的文笔既清晰明了、又富有个性。尽管《物种起源》引发了争议，但该书没有反对神的存在，而只是反对用智慧设计*的概念（即相信地球上的生命是由智慧之神创造的）来解释物种的起源。[2]很多当代科学家认为，科学和宗教不是相互对立的，因为双方解决的是两个截然不同的问题。[3]宗教试图解释的是个体生命存在的意义，并提供精神寄托；科学着力的是理解地球上的整体生命，而并不对特性或行为方式作出价值判断。

该书与人类、动物和植物研究相关[4]，所提出的新理论至今仍然能够有力地解释物种变化。[5]该理论对生物学、进化心理学*（依据进化理论研究人类的思想和行为）及植物学（对植物的研究）等的研究十分重要。该理论也或多或少地影响了神学（对宗教思想的系统研究，一般通过梳理宗教经文进行）、现代哲学、海洋学（对海洋的科学研究）、语言学（研究语言的结构和性质的学科）、人类学（对人类文化和社会生活的研究）、历史学和经济学等诸多学科。

> "人类是猿类还是天使？上帝啊！我站在天使的这边。对于那些新潮理论，我深为厌恶、愤慨之至，完全不能接受。"
>
> —— 本杰明·迪斯雷利的演讲，1864 年，引自彼得·鲍勒：
> 《查尔斯·达尔文：其人及其影响》

作者生平

查尔斯·罗伯特·达尔文于 1809 年出生在一个富裕的知识分子家庭。父亲希望他从事一份受人尊敬的职业，在他 16 岁时把他送到爱丁堡大学医学院学习。但达尔文压根儿就没打算当一名医生。他觉得医学课程枯燥无趣，看到不使用麻醉剂的手术就令他不安。相反，达尔文对自然历史（自然世界）产生了兴趣，他开始广泛地阅读、收集标本和解剖小动物。他与一位标本制作师成了朋友，从他这里学到了怎样给鸟类去毛、填充；他还结交了动物学家*罗伯特·格兰特*，罗伯特向达尔文介绍了法国生物学家让-巴蒂斯特·拉马克*提出的嬗变*理论（首次尝试解释同一物种间的变化可以代代相传的理论)。⁶达尔文还加入了布里尼学会*，这个社团里的成员都对自然历史感兴趣，他还参加了科学研究辩论会。

达尔文荒废了医学学业，惹恼了他的父亲；于是在1827年，把他送到剑桥基督学院学习神学。但是，达尔文延续着自己对自然史的兴趣。他参加了植物学的各种讲座，还和身兼牧师的植物学家约翰·史蒂文斯·亨斯洛*进行了时间跨度很长的植物收集考察活动。达尔文还对地质学*（探究地球岩石形成和历史）产生了兴趣，并且他还学习了地质学家亚当·塞奇威克*讲授的一门课程。

1831年，达尔文完成了剑桥大学的学业，登上了"贝格尔"号皇家军舰，出任随船博物学家，此外他的另一身份是船长罗伯特·菲茨罗伊的绅士客人。达尔文不得不自掏腰包，而且航游持续的时间远远超过了他预期的两年，最后用了五年才结束航程。航游期间，他收集到了无以计数的物种样本，他对一些做了解剖，另一些则做成了标本。他还花了不少时间对所看到的地质地貌进行考察，并且找到了可以支撑地质均变论*的证据，这种理论认为，地球随着时间的变迁，经历自然演变过程而发生了一系列的物理变化。

达尔文于1836年回国，在他的日记——《贝格尔号航行日记》作为游记出版后，立刻声名鹊起。由此他赢得了父亲和同龄人的肯定，并被当选为地质学会秘书。这一时期，他又与生物学家托马斯·亨利·赫胥黎*和地质学家查尔斯·莱尔建立了友谊，他们对达尔文的思想和学术生涯起到了无法估量的作用。⁷

1839年，达尔文与他的表妹艾玛·韦奇伍德*结婚。不到一年，达尔文突然病倒，夫妻俩搬到了肯特郡的达温宅。也许这个疾病正是他在寻找的一个借口，借以离开社会关系的烦扰，潜心科学研究。[8]达尔文夫妇共育有十个孩子，其中三个很小就夭折了；他对家庭的重视和对自然科学的迷恋不分上下。《物种起源》出版后，由于身体状况较差，他不能参加由此产生的激烈的公开辩论，但通过与赫胥黎这样的一些密友的书信来往，他对当时的时事保持着密切的关注。在肯特郡，达尔文生活平静，很少参与社交活动，把精力全部放在家庭生活、著书及撰写科学论文上。

创作背景

在青年时代，达尔文就开始收集各种矿物、鸟蛋和昆虫——这在他作为科学家和医学家的父亲罗伯特·沃林·达尔文 * 的眼里，无异于自我放纵、浪费时间。家族里还有其他几位科学家，其中包括达尔文的祖父伊拉斯谟斯·达尔文 *，他是一位医生和博物学家；达尔文的表兄弗朗西斯·高尔顿 * 是位统计学家和人类学家。在母亲的家族中，外祖父乔西亚·韦奇伍德 * 是当时著名的陶艺企业家，同时也是杰出的废奴主义者 *，是废除奴隶制斗争中的积极分子。这两大家族的成员基本都是一神论者（属于基督教信仰两大分支之一新教的一派）。达尔文厌恶奴隶制，支持《1832 年改革法案》*（该法案通过立法授予许多之前被剥夺公民权之男性公民以选举的权利），这些都表明了他对自由主义 * 和改革思想的基本倾向。

由于经济上拥有足够的保障，达尔文比起一般维多利亚时代的普通人拥有更多的闲暇时间进行科学探索，追求其对自然历史的兴趣。他的社会地位使他有机会结识了那个时代最伟大的科学家、哲学家及文学家，[9]包括既是科学家又是数学家的约翰·卢伯克 *，以及社会理论家哈里特·马蒂诺。在这种环境下，达尔文的理论不断发展，他在植物学、动物学（关于动物的研究）以及生物遗传*（亲代可以把各种特征遗传给子代）中的实践支持了这些理论。1859年，在《物种起源》出版之际，达尔文就获得了查尔斯·莱尔、托马斯·赫胥黎和植物学家约瑟夫·道尔顿·胡克*提供的在专业与个人方面进一步的支持。尽管达尔文的进化论招致了宗教争论，[10]但他直到生命的尽头一直保持着专业学术上的名声。

1. 爱德华·J.拉尔森：《进化：科学理论的卓越历史》，纽约：现代图书馆出版社，2004年。
2. 马克·雷德利：《如何阅读达尔文》，伦敦：格兰塔图书公司，2006年。
3. 史蒂夫·琼斯：《达尔文岛：英格兰花园中的加拉帕戈斯群岛》，伦敦：利特尔和布朗出版社，2009年。
4. 辛西娅·德尔加多："寻找医学进化"，《美国国家卫生研究院记录》第58卷，2006年第15期，第1—8页。
5. H.艾伦·奥尔："用遗传学检测自然选择"，《科学美国人》第300卷，2009年第1期，第44页。
6. 珍妮特·布朗：《查尔斯·达尔文传（第一卷）：航游》，普林斯顿：普林斯顿大学出版社，1996年，第75—76页。
7. 布朗：《航游》，第355—356页。
8. 珍妮特·布朗：《查尔斯·达尔文传（第二卷）：地方的力量》，伦敦：皮米里科出版社，2003年，第5页。
9. 布朗：《地方的力量》，第5页。
10. 布朗：《地方的力量》，第5页。

2 学术背景

要点 🔑

- 自然科学涉及如何理解物质世界的问题。

- 在 17、18 世纪时期所出现的启蒙运动中，现实社会和学术界转向理性，一些科学家对新物种发展也已经开始了思考，但还无法就其产生的机制做出完整的解释。

- 《物种起源》一书提出了"自然选择"机制，回答了物种如何随时间逐渐演变这一当时非常紧迫的问题。

著作语境

没有哪本书比查尔斯·达尔文所著的《物种起源》（1859）更能恰当地体现维多利亚时代的特点了。19 世纪的英国社会新思潮迭出，社会革新此起彼伏，国家政府和教会都面临着挑战，其中有《1832 年改革法案》，该法案授予之前数以千计被剥夺选举权的男性公民以选举的权利；以及关于人在整个自然界中的地位的辩论和废奴（反奴隶制）运动。这些新生的社会科学思潮、非国教派 * 的宗教教派（抵制英国国教的宗教信仰）以及反酒精禁酒运动 * 共同对后启蒙时代的社会以及主流意识形态发起了挑战。

这个新旧交锋的前线，早已听到隆隆的炮声。欧洲及美国呼吁所有男人都应该享有投票权（男性普选权），同时，自由主义也在 17 世纪中期诞生（通过针对政府进行的制衡与限制系统来保护个人的自由不受侵犯）。《物种起源》里所论述的那些颠覆性的思想揭示了生命是由自然而非上帝创造的。这对最高级别的权威提出了挑战。这一著作有力地质疑了主导范式*（获取和理解知识的模式）——即认为生物是静止的、永不改变，它们由神创造，而神又赋予人类对生物的统治权。虽然曾经有许多科学家一直想要挑战这种模式——但勇敢者寥寥。

> "当我们把每一种复杂的构造和生命本能看作是有利于生物体本身的许多精巧设计的综合累积，类似于所有伟大的机械发明……当我们用这种方式观察每一种有机生命的时候，从我自身的体验来说，博物学的研究将变得更为有趣！"
>
> —— 查尔斯·达尔文：《论依据自然选择的物种起源》

学科概览

根据基督教的教义，上帝引来洪水惩罚人类，由此造成的伤害形成了地球。地球是个静态的废墟，从此再没有发生过改变。今天的物种自古有之，它们的灭绝或改变是不可想象的。

基督教教义并不是唯一重要的思想上的影响。古希腊思想家柏拉图*的哲学也有类似的影响。柏拉图认为，宇宙包含了所见与所感（或者实际上想象得到）的万事万物固定不变的理想类型，而这些理想类型在我们日常生活中隐匿不见，只有扭曲了的变异形式才可见。当这种被称为本质主义*的思想应用于自然环境，生物的变化就变得不可能了。例如，野生大象的身高彼此都可能有所不同，但每只大象的内在都隐藏着一幅理想大象的蓝图，[1]"变化"不过是一种变异形式而已。

这一观点早在18世纪就遭受质疑，当时地质学家开始发掘出某些贝壳类动物，这些动物很久以前存在过，但后来却绝迹了；这一发现对圣经描述的世界构成了挑战。这些科学家们又陆续发现了更多已经灭绝的生物形式。为了从正统宗教的角度解释这一地质学的发现，基督教徒认为它们一定是间歇性灾难的结果：史前的洪水不仅仅是一次，而是多次，最后一次便是诺亚*所经历的那一次。

到 18 世纪末，其他学科也出现了支持进化论的证据。在《物种起源》发表的前一年，美国新泽西州发现了一具几乎接近完整的鸭嘴龙的恐龙骨架。再向前追溯十年，1841 年，英国博物学家理查德·欧文*将恐龙归为一个单独的分类学*组群（动物特有的目），并创造了恐龙类这个名字。事实上，达尔文没有把恐龙列入自己的著作中，因为当时发现的恐龙标本非常少；[2] 人们对恐龙的情况也还一无所知，也不知道恐龙的后代是不是还存在。比较解剖学*（即对不同物种解剖学的比较与对照的研究）为进化论提供了更加强有力的证据。

许多脊椎动物，即有脊椎的动物，具有非常相似的解剖结构。所以，人类手掌中五个手指的结构在蝙蝠的翅膀中也能观察到，蝙蝠在其解剖中反映出五个（"手趾"）的结构。此外，比较解剖学家观察到，人类胚胎的发育阶段与鸟类、爬行动物和哺乳动物的相同。所有这些证据都与智慧设计的概念（即地球上所有的生物是由神圣的"设计者"所设计的）相矛盾，由此达尔文开始将它视为生物有共同祖先的支持性证据。

学术渊源

众多学者对达尔文都产生过影响，他们的观点也出现在《物种起源》中，特别是地质学家查尔斯·莱尔，他发展了均变论。该理论认为地球在漫长的地质演变时期，经历了一系列的物理变化等自然演变过程。达尔文认为，如果物理环境可以改变，那么动植物也可以改变。事实上，动植物必须适应已经变化了的环境，否则就会灭绝。

法国生物学家让-巴蒂斯特·拉马克对达尔文也至为关键。他深信存在巨链 * 这一学说，即自然物种从最简单到最复杂是一种有序的排列。拉马克认为这一想法是物种升级的阶梯，物种努力变得更加复杂。他提出了一个理论：个体物种可以通过调适自身的身体结构以适应环境问题，并将调适后所形成的有益特征遗传给后代。[3]

另外一个主要的影响来自具有开拓精神的统计学家托马斯·罗伯特·马尔萨斯 *。他于 1798 年首次出版了《人口论》。该书启发了达尔文，让他思考存在抑制人口过剩的自然机制的可能性。

如果没有博物学家、解剖学家 *（动物解剖学专家）和分类学家（生物分类专家）的工作，达尔文的自然选择学说就不会存在；这些专家帮助达尔文对在"贝格尔"号航游期间所搜集到的大量标本进行编目和文字说明。解剖学家理查德·欧文和鸟类学家约翰·古尔德（鸟类专家）对一些标本进行了描述。古尔德意识到从加拉帕戈斯群岛带回来的鸟类，不是达尔文所认为的黑鹂、蜡嘴鸟和燕雀，实际上是 12 种不同种类的雀科鸣鸟。[4]

---

1. 爱德华·J. 拉尔森：《进化：科学理论的卓越历史》，纽约：现代图书馆出版社，2004 年。

2. 黛博拉·吉百利:《可怕的蜥蜴：第一批恐龙猎人和新科学的诞生》，纽约：亨利·霍尔特出版社，2000 年。

3. 丽贝卡·斯托特:《达尔文的幽灵：寻找第一个进化论者》，伦敦：布卢姆斯伯里出版社，2012 年，第 194—197 页。

4. 阿德里安·德斯蒙德和詹姆斯·摩尔:《达尔文》，伦敦：迈克尔·约瑟夫出版社，1991 年，第 209 页。

3 主导命题

要点 🔑

- 根据所处的特定环境和所遗传的有机组织，生物发生改变以适应自然环境。

- 圣经信奉者（那些从字面上理解圣经的人）对此不予认同；一些深受尊重的科学家认同他们的观点。

- 达尔文指出，物种的确会随着时间而变化，他还着手寻找产生这种变化的机制。

核心问题

查尔斯·达尔文所著《论依据自然选择的物种起源》解释了地球上的所有生物是如何通过自然选择过程而进化的。

达尔文研究的核心问题是为什么一些动植物长期存在，而其他动植物却归于灭绝，为什么许多灭绝动植物和现存物种之间存在相似之处。

在 19 世纪，人们普遍认为世界和栖息生物一直以来都是一成不变的，都是由上帝创造的。达尔文搭乘贝格尔号进行了世界航游，在南美洲地质研究中看到了第一手的证据和化石。这些与传统认知存在着矛盾，他开始质疑当时的学说。到 1837 年，他确信生物是经过演变而来的，并且他想知道这些进化改变是如何产生的。久而久之，他开始坚信新物种的产生并非神灵干预的结果，主要原因是为了适应环境的变化。[1]

> "体系的一部分对另一部分，以及对生活条件的一切巧妙适应，还有一个生物对另一生物的巧妙适应，这些是如何完善的？"
>
> —— 查尔斯·达尔文:《论依据自然选择的物种起源》

参与者

基督教教义声称，自从上帝创世纪以来，这个世界就从未改变；这就给地球确定了 6 000 年的年龄。如果地球从来不曾发生过物理变化，生物也就不需要改变了。然而，法国博物学家乔治-路易斯·勒克莱尔（后名为孔蒂·德·布封*）在 1749 年对此提出质疑，认为各种生命都有其自身的历史，地球的历史也远不止 6 000 年。[2] 布封通过观察还发现人类和其他灵长类动物，如黑猩猩，之间有着相似之处，并提出人类和其他类猿拥有共同祖先的观点。尽管布封提出了生物变化的观点，但他未能形成一个合理连贯的机制来阐释为何这些变化会产生的原因。[3]

博物学家让-巴蒂斯特·拉马克在《动物学哲学》（1809）中对圣经提出了另一个挑战性的理论：基于物种突变的进化论。该理论认为是环境挑战迫使物种在相当长的时间内改变自身结构以获得个体优势。此外，他还认为后代可以从亲代那里继承这些修正之处。[4]1826 年，英国博物学家罗伯特·格兰特开始公开谈论涉及进化的观点，并在英国提倡拉马克关于物种的嬗变理论。[5]

针对神创论*面临的这些挑战，威廉·佩利主教*认为各种适应性的变化都是由超自然的上帝所创造的。解剖学家乔治·居维叶*也对拉马克的观点进行批判，坚持认为物种是不可改变的（不会改变，也不可改变）。然而，他排斥进化论并不是因为宗教信仰上的原因，而是出于证据方面的原因——化石记录未能展现出物种主动进化的状态。[6] 在居维叶看来，灾难性的变化出现之后紧跟的就是物种自发的繁殖行为。

当时的论战

在 18 世纪，地质学家开始发现一些包含先前灭绝事件记录的岩石（灭绝事件时期，大量物种同时死亡）。这个时期的科学家为了迎合基督教教义引入间歇性灾难理论，理论假设每次灾难发生后，上帝就重塑地球上的生物。最后一次的灾难就是诺亚遭受的洪水。然而，此种解释并不能说明为什么经历了如此浩劫，有些动物物种遭遇灭绝，而其他动物物种却能幸存下来，例如，中新世时期的獾类物种与我们现在的獾几乎完全相同。[7]

以在古代岩石中发现的化石为证据，苏格兰地质学家詹姆斯·哈顿*在 1788 年提出了均变理论。该理论认为，地球在过去经历了持续的物理变化，同样的变化过程也一直延续到了今天。这种变化持续不断，渐进渐变，而非一系列灾难性事件。哈顿的理论当时被广泛忽视，直到 19 世纪，才引起了人们的注意，当时的地质学家查尔斯·莱尔设法将它完善并加以推广。[8]

法国生物学家让-巴蒂斯特·拉马克在他的《动物学哲学》（1809）中提出了一个有力的论点。拉马克吸收了希腊哲学家柏拉图思想中的本质主义观点，该观点认为，每一种动物或是物理客体都拥有一套身份和功能必需的属性特征，以及存在巨链的观点（认为所有生物按结构分层，且高优低劣），他认为所有生物体都不断力求变得更加复杂，最终目标就是变得像人类那样复杂。随着一些生物体结构变得越来越复杂，在结构等级底部就会出现缺口，这些缺口由简单而可以自发生成的生物体所占据。拉马克认为，两种力量引导着这一过程：趋于复杂化的内在驱动力和环境。他还认为，生物会通过调整自身构造来应对生态挑战，并将这些变化遗传给下一代。这种论点颇具说服力，一直是最具影响力的进化理论，直到达尔文和威尔士博物学家阿尔弗雷德·拉塞尔·华莱士提出类似的结论。[9]

然而，拉马克有关物种嬗变的理论与启蒙运动中的激进唯物主义相关（这里的"唯物主义"是指所有的物理现象都有其物理成因的假设），因而受到敌视。达尔文年轻时就读过佩利主教的《自然神学》（1809），这本书的部分章节就回应了这一理论。佩利在书中称适应是上帝超能力的创造结果。[10]的确，自然界中这种适应的存在为上帝的存在提供了主要的一个哲学论据，这便是众所周知的"天意设计"的论据。

莱尔也在《地质学原理》（1830，1833）中对拉马克的理论进行批判。尽管莱尔撰写了无机物的均衡变化，但他拒绝相信生物变化的可能性。相反，他提到每个物种都有其"创造中心"，是为某个特定的环境而设计；当这个生存的环境发生变化，物种也即消亡。[11]

---

1. 爱德华·J. 拉尔森：《进化：科学理论的卓越历史》，纽约：现代图书馆出版社，2004 年。

2. 丽贝卡·斯托特：《达尔文的幽灵：寻找第一个进化论者》，伦敦：布卢姆斯伯里出版社，2012年。

3. 乔治-路易斯·勒克莱尔（孔蒂·德·布封）："大自然的时期"，《自然史》，巴黎：国家出版社，1749—1788年。

4. J. B. 拉马克：《动物学哲学：关于动物自然史的论述》，休·艾略特译，芝加哥：芝加哥大学出版社，1984年。

5. 斯托特：《达尔文的幽灵》。

6. G. 居维叶：《动物自然历史元素表》，巴黎：鲍杜因出版社，1798年，登录时间2016年2月16日，https://archive.org/details/tableaulment00cuvi，第71页。

7. 罗伯特·钱伯斯：《自然创造史的遗迹》，伦敦：约翰·丘吉尔出版社，1844年。

8. 拉尔森：《进化》。

9. 拉尔森：《进化》。

10. 威廉·佩利：《自然神学或神的存在和属性的证据》，伦敦：J.福尔德出版社，1809年。

11. 查尔斯·莱尔：《地质学原理》，第二卷，伦敦：穆雷出版社，1830—1833年，第二章。

4 作者贡献

要点 🔑

- 科学家长久以来一直在寻找用来解释物种内部和物种间的变异现象的工作模式。最终达尔文（以及和他同时代的阿尔弗雷德·拉塞尔·华莱士）找到了这个模式。

- 自然选择为解释不同物种在漫漫时间长河中是如何形成、如何变化的提供了工作机制。

- 达尔文提出，生物个体的特征各不相同，彼此间为了在资源有限的世界中生存而不断斗争。

作者目标

　　查尔斯·达尔文撰写《论依据自然选择的物种起源》一书的目的就是解释适应性和进化过程中的变化。这种理论需要对地球上生物的多样性作出解释，还要解释为什么生物体会展现出这样的外观和行为。适应包括机体局部的物理变化，例如翅膀，还包括某种特定的行为习惯——二者都旨在提高生物生存和繁殖的机会。这种适应不可能偶然出现，所以需要我们对之作出解释。[1]

　　自古希腊时代以来，进化这一概念就一直存在。在达尔文发表《物种起源》之际，许多哲学家和科学家都相信动物和植物的生命随着时间的推移而进化。[2] 到了 19 世纪，地质学、胚胎学和解剖学积累了大量有利于进化的证据（"胚胎学"是指对在动物子宫中胚胎发育的研究）。科学家曾经尝试解释动物和植物的各种形态和类型是如何形成的，但没人能给出令人信服的答案。达尔文改变了这种局面，他提出的合乎逻辑的论点对科学家和一般读者来说都易于理解。

> "人类的起源和历史的问题将会被阐明。"
> —— 查尔斯·达尔文:《论依据自然选择的物种起源》

研究方法

在《物种起源》一书中，达尔文使用比较法构建了科学的论证：他向读者展示了一组详细的观察结果，然后用推理以论证自己的观点。例如，为了展示多样性是如何在自然界中产生的，他用家养动物作为例证进行了说明。他还解释了狗的选择性繁殖如何导致品貌上各不相同的结果，尽管它们都是一个共同祖先的后代。这表明：少数几个品种就可以产生大量品种。达尔文由此得出结论：这种情况在野生动物上同样适用。

达尔文也为他的渐进主义观点提供了证据。以眼睛为例，他指出：哺乳动物*的眼睛一开始有可能也只是简单的光敏器官，和无脊椎动物（没有脊椎的动物）的类似。[3] 当然，因为眼睛由软组织组成，无法出现在化石中，但达尔文还是原则上坚持自己的观点，推断出人眼形成的进化阶段。通过这种方式，他反对智慧设计这一概念，而运用自然选择取而代之。[4]

时代贡献

达尔文和同时代的博物学家阿尔弗雷德·拉塞尔·华莱士利用从地质学、胚胎学和解剖学中获得的证据，构想出了一个关于进化的原创理论。这与让-巴蒂斯特·拉马克和其他一些德国科学家所用到的方法大相径庭，他们的方法是推测和理论推导。值得注意的是，达尔文和华莱士在创立自然选择学说时都是从人类出发，而最终又回到动物上去。

虽然达尔文和华莱士两人被认为共同发现了自然选择规律，但仔细阅读他们俩在 1858 年联合发表的文章，就会发现他们对这一理论有不同的态度。只有达尔文认为同一种群中各成员间的竞争是最激烈的，远远大于不同物种间的竞争。达尔文还指出，拥有较近共同祖先的物种往往看起来就较相似，而拥有遥远的共同祖先物种间的共同特征则较少。[5] 这种认识后来发展成为趋异 * 原则，和物种随着时间推移是如何进化的结论。

另一方面，在物种形成的过程中，华莱士强调的是环境的重要性。[6] 他认为，食物供应和猎食能力对物种数量增长具有特别巨大的影响，并得出结论："每年死亡的数字必然巨大，由于每只动物的个体存在依赖于自身因素，死亡的必定是那些最为羸弱的——无非是些太过幼小、极其老迈或身患疾病的个体，而那些能够延长生命的只能是最为健康而又最富活力的个体：即那些平常最精于获取食物，并又善于躲避众多天敌的个体。"

与达尔文相反，华莱士强调，个体的*弱小*并不是基于生物遗传，而只是纯属偶然；他给出的例子是那些过于年迈或是幼小年轻的动物。

---

1. 马克·雷德利：《如何阅读达尔文》，伦敦：格兰塔图书公司，2006 年。

2. 爱德华·J.拉尔森:《进化:科学理论的卓越历史》,纽约:现代图书馆出版社,2004 年。

3. 雷德利:《如何阅读达尔文》。

4. 雷德利:《如何阅读达尔文》。

5. 迈克尔·布尔默:"皇家学会会员阿尔弗雷德·拉塞尔·华莱士的自然选择学说",《皇家历史科学学会杂志》第 59 卷,2005 年第 2 期,第 125—136 页。

6. 布尔默:"自然选择学说",第 125—136 页。

第二部分：学术思想

5 思想主脉

要点 &

- 达尔文所探讨的主题是任何群体中的个体都各不相同。
- 这种差异导致了存活率和繁殖率的不同。
- 拥有最适于生存的特质的个体能够存活，进而繁衍后代；并将这些特质遗传给后代。

核心主题

查尔斯·达尔文所著《论依据自然选择的物种起源》一书的核心主题就是变异、生物遗传和种群内部的资源竞争。这些共同构成了进化机制中的重要内容，达尔文将之称为自然选择。

生物学有两种变异。第一种是由于遗传差异（即决定生物体特性的基因 * 差异）引起细胞、生物体个体或生物体群体的变异。这也就是所谓的"基因型变异"。第二种变异考虑到了环境对遗传潜能的影响，称为"表型 * 变异"。变异既可以体现在身体外观上，例如身高，也可体现在行为方式上。"生物遗传"用来描述基因从父母（或祖先）传递给后代的过程。

最后，生物学中的"竞争"指的是两种或两种以上生物之间的相互作用，这些生物体都需要获取相同有限的资源（例如食物、水或配偶，有了这些资源，生物才能生长、存活和繁殖）。如果一个生物消耗或捍卫了某种资源，别的生物体就无法获得此资源，因此，竞争对手减少了对方生长、繁殖和存活的概率。

> "我们在这里不是为了关注自己的希望或恐惧，只关注尽己所能可以探寻到的真理。"
>
> —— 查尔斯·达尔文：《人类的由来及性选择》

思想探究

在《物种起源》的开头，达尔文就用家养的动物和植物这些日常生物为例，阐释物种内部存在着固有变异。通过观察，他发现一些家畜具有一种"非凡的变异倾向性"，[1] 尤其是犬类。他用了不短的篇幅，描述了多种犬类品种及其形体差异，并做出推测：它们可能由少数野生物种演化而来。达尔文将之称为种类，是因为它们可以与家犬（犬科）进行杂交繁殖，因而可以将它们归为一个物种。从家养动物出发，他进一步对发生在野生动物和植物中的自然变异进行了探索。他指出，家养犬类是人为操纵的产物——具体使用的是选择性育种的方式。根据这一方式，育种员选择某些特征明显的动物，如体型或智力等，进行配种，进而发展这些特征加以遗传，这样一来，这种家养的犬类就可能不是自然地产生的。但他强调了一个事实，即犬类品种中观察到的变异肯定是其野生祖先中与生俱来的特性。[2]

达尔文讲到的第二个主题是"为生存而奋斗"。有些个体比别的个体更擅长掠夺资源。达尔文推断，生存竞争中最成功的个体都会拥有某种或某些特性变异后产生的适应能力，为这些个体提供优势地位。这些适应能力更强的个体会比其他个体繁殖数量更多的后代。后代通过生物遗传的方式继承父母的优势特质。但达尔文未能描述所涉及的真正机制，因为那时候基因（将特征代代相传的生物材料）尚未被发现，尽管如此，他仍认为这种自然选择过程会使得某个种群中更多的个体在特定时期更能够适应其环境。值得注意的是，达尔文用到了"适者生存"*这个词作为比喻，说明并非所有的个体都会为下一代做出同样的贡献。"适者"这个词并不含价值判断：在生物学中，这个词用于指代那些存活下来的个体，更为重要的是指那些有后代的个体。

达尔文利用选择性育种提出了渐进性这一主题，展现通过连续几代递增的细小变化，适应就会出现。他承认，家养动物的变异可能对它们本身没有好处，而是为了"人类自己的使用需要或一时兴起"。[3] 尽管这些不是真正的适应，但确实可以说明生物有能力经过漫长的时期发生渐变，这一点不仅可以证实渐变论这一概念的正确性，并且可以证实个体是可以从亲代那里获得遗传特征的。

语言表述

《物种起源》这本书的写作对象是受过良好教育并具有批判性思维的普通读者。读者并非必须具备科学背景才能理解该书，因为该书语言风格与当时的标准散文并无二致。

达尔文在讲述自己的发现时，是拥护进化论的，含蓄地排斥神创论（这种学说认为圣经关于世界与动物的创造的解释是完全正确的）。达尔文用的术语是"遗传性变异"，而不是"进化"。"进化"这一术语在 19 世纪后半叶才得到更广泛的应用，仅在1872 年第六版中出现，而"进化的"一词仅在作品的最后一个句子中使用过一次。[4]

达尔文使用"自然选择"这一术语来解释生物是如何变化及适应的。他只有几次直接提到"智慧设计"这一概念，写作内容完全从物理性质和自然规律的角度描述生物本质。他总是力求避免正面陈述他的理论如何颠覆宗教对生命起源的描述。然而，在他的读者看来，他的观点相当清晰，而达尔文有争议性的观点，可能促成了实证主义。实证主义认为从感官经验所得来的信息，经过理性和逻辑的阐释，才可以成为权威知识的唯一来源。

达尔文全书都使用了类比的写作手法，例如对家畜和野生动物的比较。而使读者能够轻而易举理解他的，不仅仅是因为他提到了家养动物，而且还提到了家庭生活。[5] 一个典型的例子就是他描述如何用茶杯从本地的一个池塘舀出泥浆。他用这样的方式非正式地表达他的观点，对维多利亚时代的读者来说易于理解。他的目的也是为了获取更广泛的读者群："我有时候认为，科普性的论著与原创作品对于科学的进步几乎有着同样重要的作用。"[6] 达尔文成功了：《物种起源》现在已是世界名著。

1. 查尔斯·达尔文：《论依据自然选择即在生存斗争中保持优良族的物种起源》，吉利安·比尔作前言及注释（1996 年，2008 年），牛津：牛津大学出版社，1860 年。

2. 达尔文：《物种起源》。

3. 达尔文：《物种起源》。

4. 马克·雷德利：《如何阅读达尔文》，伦敦：格兰塔图书公司，2006 年。

5. 乔治·列文：《作家达尔文》，牛津：牛津大学出版社，2011 年。

6. 史蒂夫·琼斯："查尔斯·达尔文给托马斯·赫胥黎的一封信"（1865 年），《达尔文岛：英格兰花园中的加拉帕戈斯群岛》，伦敦：利特尔和布朗出版社，2009 年。

6 思想支脉

要点 🔑

- 达尔文的思想支脉还包括性别选择（通过选择伴侣而产生进化的过程）、物种的渐变和物种形成（不同的物种是如何形成的）。

- 这些思想支脉，特别是性别选择和物种形成理论，与当代进化的研究依然息息相关。

- 这些思想中最重要的是性别选择，即使在现代进化理论体系中，仍旧是自然选择中的重要因素。

其他思想

查尔斯·达尔文的《论依据自然选择的物种起源》中的另一个重要思想支脉就是自然选择中被称为"性别选择"的这个部分。这个理论部分解释了一个悖论：尽管个体的某些特征会降低生存机会，但却依然在进化中得以保留，如孔雀的尾巴。

另一个重要思想支脉是关于新物种的起源问题——物种形成。达尔文依据自然选择的进化论需要大量微小却有利的改变缓慢且渐进的积累，就是所谓的渐进性。[1] 在该书第六章中，他特别提到了一个问题："首先，如果物种（在漫长的时期）通过细微不易察觉的变化从其他祖先物种演变而来，但为什么我们没有在任何地方看到众多的过渡形态呢？"[2] 达尔文在这里提出了化石记录中缺少过渡形态的问题，他接着长篇讨论了自然选择中必然有方法能够促成不同物种的产生。

> "性别斗争可分两种：一种是同性个体间的斗争，一般是雄体为了赶走或杀死自己的竞争对手，雌性表现得并不主动；另一种同样也是同性个体间的斗争，但目的则是为了激发或吸引异性——一般是雌性——的注意，这时候的雌性再不会表现被动，而是会选择更中意的伙伴。"
>
> —— 查尔斯·达尔文：《人类的由来及性选择》

观点探究

性别选择有两种形式：同性选择，即同性成员为获得交配机会而相互竞争，如雄红鹿使用鹿角相互撞击以夺得交配权；和异性选择，即异性成员被对方的某些特征所吸引，如雄性孔雀夸张多彩的尾羽。如果雌性孔雀所选中的雄性孔雀与其竞争对手具有不同的基因特征，自然选择则就此出现。[3]

达尔文认为，自然选择的演化是一个渐进的过程。新物种是经历长时间缓慢且不断累积的变化过程而形成的。对达尔文来说，一个主要的难点是在化石记录中找不到这些变化的证据，即找不到以中间状态形式出现的化石形态。

即便是今天，科学家也没能找到许多记录中间形态的化石。进化论生物学家史蒂芬·杰·古尔德*和尼尔斯·埃尔德雷奇为这一空白提供了一个解释：这些中间形态是真实存在的，并且代表了一个稳定期，因为物种在几百万年间并未发生多大的变化。他们认为紧跟在这些稳定期之后的就是迅速变动期，导致了新物种的产生，这就是所谓的"间断平衡论"。[4]

根据这一理论，产生新物种的变化通常不是来自主流种群中某些细微而缓慢的变化，而是出现在该种群中的一个小子集中；例如生活在栖息地边缘或孤立的一个小群体中。当环境条件发生变化时，由于该群体规模较小，这些"外围"或"地处边缘"的生物就会经历激烈的选择过程，并快速变化。因为它们的数量相对较少，所处的地理位置也很偏远，就没有留下显示中间阶段状态的化石。这些成功变异的新物种接下来就可以遍布于它们祖先所活动的地理区域了。[5]

事实上，达尔文登上贝格尔号航船，在加拉帕戈斯群岛航行时所收集到的地雀，每一只都是由大陆上的雀类进化而来，目的是为更好地适应太平洋偏远岛屿上特定的栖息地环境。间断平衡理论并不意味着进化只能发生在快速激烈的变动过程中。科学家也观察到了确实存在着渐进的演变——特别是对我们人类自身而言。[6]

物种形成是否意味着生物体以确定的类型存在？达尔文对独特物种的存在持保留态度："在此，我也无意讨论物种的各种定义。没有哪一种定义是所有博物学家都能认同的；但每一位博物学家在讲到物种时又都隐约地知道所要表达的含义。"[7] 在 20 世纪下半叶，博物学家们曾提出 20 多种不同的对"物种"的定义。[8] 有人认为，如此多不同的定义本身实际上已经证明了独特的物种是不存在的，并且他们倾向于将生物体视为渐变整体中的一个环节，而非独特的实体。他们还说，将物种进行分类的这种做法，更多是人类思维的产物，为的是能够简单地对事物分类，而不是真实直观地去反映自然。[9]

被忽视之处

尽管性别选择理论对进化论来说十分重要，但在很大程度上被忽视了近一个世纪，原因是人们很难理解动物是如何才能显示出性别选择的。达尔文本人对这个概念也存有疑问，因为在当时，所谓的动物是被视为机器人，完全屈从于本能的摆布。因此，尽管科学家现在明白这种选择作为一种机制客观存在，[10] 但在达尔文时代，做出这样思想上的跨越十分困难。

如今，性别选择已成为现代进化生物学和行为生态学 *（用动物所处的环境来解释其行为的学科）的核心内容。这在某种程度上应归功于英国统计学家罗纳德·费舍尔 *，他在 1915 年建立了一个性别选择模型。他认为，某些特征本身（如雄性孔雀的尾巴）是能够自我进化的，前提是这一特征本身（如雄孔雀的尾巴）及过度反映该性状的偏好取向都有遗传基础。如果雌性携带了某种偏好基因，它们的雄性子代就会获得一种基因，将这种偏好特征加以发挥，这会导致具有该特征及偏好的个体数量比例增加。[11] 费舍尔的著作直到 20 世纪 70 年代才引起人们的注意，因为那时候有人发现在雄性特征形成过程中，雌性的选择确实发挥着非常强大的影响。[12] 在 20 世纪 70 年代和 80 年代，还出现了不少其他的模型，均可参考。[13]

---

1. 查尔斯·达尔文：《论依据自然选择即在生存斗争中保持优良族的物种起源》，伦敦：约翰·穆雷出版社，1859 年，第 103 页。

2. 达尔文：《物种起源》。

3. J. R. 克雷伯和 A. 卡塞尔尼克："决策制定"，《行为生态学：进化论方法》，牛津：布莱克威尔科学技术出版公司出版社，1991 年，第 105—136 页。

4. 尼尔斯·埃尔德雷奇和 S. J. 古尔德："间断平衡：代替种系发生渐进主义"，《古生物学模型》，T. J. M. 肖夫编，旧金山：弗里曼库珀出版社，1972 年，第 82—115 页。

5. 古尔德，"间断平衡"，第 82—115 页。

6. W. A. 哈维兰和 G. W. 克劳福德：《人类进化与史前史》，马萨诸塞州坎布里奇：哈佛大学出版社，2002 年。

7. 达尔文：《物种起源》。

8. J. A. 马利特："现代合成的物种定义"，《生态学与进化论趋势》第 10 卷，1995 年，第 294—299 页。

9. 丹尼尔·埃尔斯坦："物种的社会建构：物种在道德上是否相关？"，《动物研究批判杂志》第 1 卷，2003 年第 1 期：第 53—71 页。

10. 克雷伯和卡塞尔尼克："决策制定"，第 105—136 页。

11. 马尔特·B. 安德森：《性别选择：行为与生态学专题论文集》，普林斯顿：普林斯顿大学出版社，1994 年。

12. 安德森：《性别选择》。

13. 安德森：《性别选择》。

7 历史成就

要点 ⚷⊷

- 达尔文是第一个成功提出物种变化机制的人。他的理论在今天仍然是最重要的范式（概念模型）。

- 达尔文专注物种起源研究 21 年，对自然选择做出了最充分的论证。《物种起源》还一直影响着包括生命科学在内的许多其他学科的发展。

- 达尔文不了解通过生物物质（现在被称为基因，是直到 1906 年才建立起来的一种研究领域）的遗传机制，这意味着他错误地理解了遗传特性是如何代代相传的。

观点评价

查尔斯·达尔文在《论依据自然选择的物种起源》一书中提出了一个重要的论点：自然选择是进化背后的驱动力。由于在化石记录中缺乏进化的确切证据，达尔文使用共同的祖先来证明他的论点，[1] 指出活动范围相隔不远的动物往往有着共同的祖先这个事实。[2] 另外，他还引入了渐变论这一概念用于生物的研究。[3]

达尔文知道他的书一出版就会引发论战，于是，他用了整整一章的篇幅对有可能出现争论的问题进行了说明，并且将该节标题定为"理论难点"。

《物种起源》在 1859 年一出版，查尔斯·莱尔、托马斯·亨利·赫胥黎和植物学家约瑟夫·胡克等就给予达尔文专业和个人方面的支持。这些宝贵朋友的帮助奠定了达尔文研究先驱*的地位，承认是达尔文的工作而非其他人的向前迈出了最重要的一步，因此荣誉理应献给他的主要研究成果。

这在 1858 年相当关键，因为在当时，博物学家阿尔弗雷德·拉塞尔·华莱士也提出了自然选择这一理论，认为这种机制导致了物种变化。有人担心，达尔文的领先地位可能会输给这个年轻人。后来，对于《物种起源》突然间在国际上引发的论战，莱尔和赫胥黎订立了攻守同盟。尽管达尔文的自然选择学说在某些方面受到广泛批评（遗传机制尚且不能为人理解；还有关于地球年龄的问题），[4] 以及对正统基督教思想发起的巨大挑战，他始终保持着卓越的学术声誉，后又相继出版了关于进化论的三部力作。

> "它将会和'牛顿定律'一样长久……达尔文为这个世界创造了一门新科学，我认为他的名字当写在所有哲学家之前。"
>
> —— 阿尔弗雷德·拉塞尔·华莱士：华莱士信函

当时的成就

当然，达尔文并不是第一个提出进化思想的人，自从启蒙运动以来，进化就一直是人们关注的问题。[5] 达尔文的祖父伊拉斯谟斯·达尔文针对这个问题还发表过几篇关于自然历史和所有生命形态之间关联的文章，包括一首名诗"社会起源"（1803）。另一个著名的先驱者是法国生物学家让-巴蒂斯特·拉马克，他在 1809 年提出物种是可变而非固化的个体。拉马克的论述极具说服力，他使进化论的信念成为一种值得尊敬的立场。博物学家埃蒂安·乔弗罗伊·圣–希莱尔*和罗伯特·格兰特都认同这一观点，正如苏格兰作家罗伯特·钱伯斯 * 在他《自然创造史的痕迹》（1844）一书中指出的，所有结构复杂的生命都从简单形态中进化而来。神职人员自然反对这本书；因为该书对传统的宗教观提出了质疑，而亚当·塞奇威克等学者则因其科学依据过于而肤浅进行了批评。[6] 尽管如此，该书在普通读者中却非常流行，部分原因是它具有可读性，[7] 这本书无疑为维多利亚时期英国的民众在接受进化思想方面做足了准备。[8]

《物种起源》第一版印刷 4 200 册，出版当天即全部售罄；在达尔文一生中，又出版了五个版本。虽然钱伯斯的《自然创造史的痕迹》发行量超过了《物种起源》，[9] 但后者影响却更大，并且受到更多的推崇，部分归因于达尔文一些挚友的帮助，部分来自他个人作为著名科学家的地位。[10]

局限性

达尔文的观点有时也被用在社会政策的制定上。达尔文本人就把自然选择理论运用得特别广泛。达尔文在《人类的由来及性选择》（1871）中就这样写道："所有不能避免子女皆陷于赤贫的人都不应结婚"，这里指的就是生存的斗争，"否则他就会陷入懒惰，而且在生存斗争中，天赋卓绝的人也就未必比平庸的人更成功。"[11]

这种言论后来发展成为社会达尔文主义 *。这一理论认为，与其他物种一样，自然选择对人类也有约束力，从基因特性上讲，某些种族比其他种族更有优势。采用这个方法，类似于适者生存这样的生物概念被更加广泛地运用于经济学、政治学和社会学。社会从"原始"状态起步、逐步走向文明，这种思想最早由英国生物学家赫伯特·斯宾塞 * 提出，同时他还提出了某些社会（西方社会）文化更加优越的观点。

优生学，一种通过选择性培育胚胎来提高人类素质的科学方法，是基于这个认识发展而来的。第一个提出优生学的人正是达尔文的表弟弗朗西斯·高尔顿，那是在1883年；他还主张实施社会控制，包括对一些被判定"不适合生育"的人进行绝育处理。"不适合生育"的人，比如说身体有缺陷或精神残疾者被禁止生育的做法在不少国家得以施行，如英国、加拿大、德国、瑞典和美国等。[12]

德国哲学家弗里德里希·尼采*发出过警示：脱离了具体的背景，《物种起源》所阐释的观点可以解读为虚无主义*，也就是说，我们的存在毫无意义。这样会免除个体道德或社会责任，因为既然生命都没有意义，为什么还要珍视它？

但是，进化论不能从价值判断或喜好爱憎的角度来理解，这种警示性的观点早在1903年就由一位叫乔治·爱德华·摩尔*的英国哲学家提了出来。他在他的著作《伦理学原理》中讲道：用某些自然属性来定义"好"这一概念必定是错误的，这会犯他所定义的"自然主义谬论"*的错误。[13]

其他人认为，如果有所进化，那就一定说明是有益的——即便是资本主义*的社会经济体系，以及这种社会体系所带来对私人利润的追求都是如此。但也有人指出，"资本主义是合理的，因为它遵循适者生存和好者生存这两条原则。但是这样一来，我们就犯下了自然主义谬论，因为所谓'好'在定义中并非指自身的东西，而是与适者生存相类似的东西。"[14]

总之，将人类道德建立在自然科学所形成的法律和概念之上是很危险的。

1. 查尔斯·达尔文：《论依据自然选择即在生存斗争中保持优良族的物种起源》，吉利安·比尔作前言及注释（1996 年，2008 年），牛津：牛津大学出版社，1860 年。

2. 达尔文：《物种起源》。

3. 马克·雷德利：《如何阅读达尔文》，伦敦：格兰塔图书公司，2006 年。

4. 雷德利：《如何阅读达尔文》。

5. 爱德华·J. 拉尔森：《进化：科学理论的卓越历史》，纽约：现代图书馆出版社，2004 年。

6. 亚当·塞奇威克："遗迹评价"，《爱丁堡评论》第 82 卷，1845 年 7 月，第 1—85 页。

7. 詹姆斯·A. 塞科德：《维多利亚时代的轰动事件：自然创造史的痕迹的非凡出版、接受和秘密作者》，芝加哥：芝加哥大学出版社，2000 年。

8. 路易斯·N. 马格纳：《生命科学史》，纽约；巴塞尔：马塞尔·德克尔出版社，1994 年，第 257—316 页。

9. 马格纳：《生命科学史》。

10. 乔治·列文：《作家达尔文》，牛津：牛津大学出版社，2011 年。

11. 查尔斯·达尔文，《人类的由来及性选择》，伦敦：约翰·穆雷出版社，1871 年。

12. 丹尼斯·苏厄尔：《政治基因：达尔文的思想如何改变政治》，伦敦：皮卡多出版社，2009 年。

13. G. E. 摩尔：《伦理学原理》，剑桥：剑桥大学出版社，1993 年。

14. 朱莉娅·坦纳："自然主义谬误"，《里士满哲学期刊》第 13 卷，2006 年，第 1—6 页。

8 著作地位

要点 🗝—

- 达尔文毕生的工作就是通过研究进化问题来解释自然世界。
- 《物种起源》奠定了达尔文后期创立诸多其他相关理论的基础。
- 《物种起源》和《人类之由来》奠定了达尔文进化论的创始人的声望；他的名字已经成为进化论的一个代名词。

定位

　　查尔斯·达尔文从幼时起，就有一种探索精神，周围的自然界令他着迷。从《物种起源》中我们也可以清楚地看出，他对于圣经中解释自然、世界起源的说法他是不认同的。达尔文在 20 岁出头的时候，他阅读的范围非常广泛，包括各种不同的学科，其中许多著作赞同神在俗世中发挥着积极作用；自然神学（这种学说认为自然界纷繁复杂而又美丽，本身就证明了上帝的客观存在）在当时的科学界占据主导地位。

　　达尔文认为这些解释是不合理的。他毕生追求的目标就是对身边观察到的自然现象作出解释。在《物种起源》及其他后续出版的著作中呈现的理论诠释了他深刻详实的思想观点。

　　达尔文书稿中的核心问题在他早年就开始萌发，有可能是在乘坐贝格尔号航游结束后不久形成的。[1] 1838 年，他阅读了苏格兰解剖学家、哲学家查尔斯·贝尔＊所著的《表情解构与哲学散文集》（1824）一书，此书总结了人类的情感是独一无二的。[2]

　　从达尔文在阅读贝尔著作时所作的随书笔记中可以清楚地看

出，他并不认同这一观点。[3] 这之后 30 年里，达尔文收集证据，这些证据最终都写入了《人类与动物的情感表达》（1872）一书。在这本书中，达尔文用共同祖先（也就是进化）来解释人类和其他动物的情感表达具有相似之处。

> "只要人类与其他所有动物均被视为相互独立的生物，就会有效地阻止我们的自然本能尽可能探寻情感表达的原因……如果我们认为截然不同，但有关联的物种具有同一个祖先，那么他们的表达就会更容易理解……认为动物的生理结构和习性都会逐渐进化的人，都会用一种全新且有趣的视角来看待情感表达这个问题。"
>
> —— 查尔斯·达尔文：《人类与动物的情感表达》

整合

进化是达尔文所有主要著作中压倒一切的主题：《物种起源》（1859）、《动物和植物在家养下的变异》（1868），《人类的由来》（1871）以及《人类与动物的情感表达》（1872）。在每本书中，达尔文都解释了自然选择是如何影响人类和其他动物成为今天我们所看到的这些样子。

在《动物和植物的变异》两卷本著作中，达尔文将对遗传与变异的讨论延伸至植物与动物。他介绍了自己的泛生论*理论。这种理论认为小片信息以胚芽（一种假设性颗粒）的形式存在的在成人身体的各个部分中流动，并形成孕育后代的信息。在这本书最后一段，他还驳斥了设计进化论的观点，这要比在《物种起源》中表现得更直接。"无所不晓的造物主肯定预见到由其制定的法律可能会造成的每一种可能的结果……更有可能是这样来看：上帝作为饲养员，是否决定了我们的家畜和植物无数的变异？这里面的许多变异对人类并没有多少用处，也非有益，更多的是不利于生物自身。是否上帝决定了鸽子要有平头与尾羽的变异，为了培育出体型奇怪的凸胸鸽和扇尾鸽？……然而不管我们多么期待，让物种全部向着有利于人类的路线而进化变异都是很难实现的。"[4]

《人类的由来》对《物种起源》中的观点进行了拓展，为动植物进化理论提供了更多的证据。达尔文还提出了有关人类起源的新观点，着重讨论人类精神和道德能力方面的演变。他认为，人类和其他动物在智力上的差异，本质上是程度的差异而非种属的差异。他还对性别选择理论——自然选择的一种形式，其中交配伴侣的选择起着关键作用——进行了拓展。之前他曾在《物种起源》中描述过此观点。[5]《人类的由来》被认为既发人深思，又有些令人震惊。

证明人是从猿类祖先进化而来的进一步的证据出现在达尔文后来的《人类与动物的情感表达》一书中："人类某些表达，比如在极端恐怖状态下出现的毛发竖立，在狂怒下牙齿的外露，几乎都很难理解；除非认识到人类曾以很低级别的动物形式而存在，理解接受起来就不再困难。"[6]

显然，达尔文一生中大部分时间都在论证进化论，将此作为解释人和其他动物对环境的适应的框架。不管是他的著作，还是他那些影响巨大的观点，如人类本是动物的看法等，都彻底颠覆了人们对生命科学和人文思想的认识。

意义

19 世纪 50 年代，关于进化论的争论十分激烈。1859 年达尔文《物种起源》的出版极大地改变了这场争论，很大程度上是因为达尔文总是能够接受所有最新资讯，进而将它们整合成有条理的论证。

首先支持达尔文的是他的密友托马斯·亨利·赫胥黎和约瑟夫·道尔顿·胡克爵士，两人积极倡导进化论。他们有着自由的学术精神，曾号召学界将宗教教条从科学研究中剥离出去，并且取得了成功。他们还支持自由派圣公会运动，该组织接受进化论，抵制所有拒绝进化新思想的传统主义者。[7]

达尔文的影响力很快就超出了学术界。1896 年，坎特伯雷大主教这一职位出现了空缺，由弗雷德里克·坦普尔*接任，他是一个坚定的进化论支持者。[8]达尔文去世后，他被安葬在伦敦威斯敏斯特大教堂，距他不远的地方，安息着伟大的科学家艾萨克·牛顿*。这标志着达尔文极高的公众威望在社会上和宗教界都获得了认可。[9]

今天，自然选择学说已经成了生命科学的一个统一的概念。

1. 马克·雷德利：《如何阅读达尔文》，伦敦：格兰塔图书公司，2006 年，第 8—15 页。
2. 查尔斯·贝尔：《表情解构与哲学散文集》，蒙大拿：凯辛出版社，2008 年。
3. 雷德利：《如何阅读达尔文》。
4. 查尔斯·达尔文：《动物和植物在家养下的变异》，伦敦：约翰·穆雷出版社，1868 年。
5. 查尔斯·达尔文：《人类的由来及性选择》，伦敦：约翰·穆雷出版社，1871 年。
6. 查尔斯·达尔文：《人类与动物的情感表达》，伦敦：约翰·穆雷出版社，1872 年。
7. 爱德华·J. 拉尔森：《进化：科学理论的卓越历史》，纽约：现代图书馆出版社，

2004 年，第 79—111 页。

8. 菲利普·基彻：《与达尔文生活：进化，设计和信仰的未来》，纽约；牛津：牛津大学出版社，2007 年。

9. 珍妮特·布朗：《查尔斯·达尔文传（第二卷）：地方的力量》，伦敦：皮米里科出版社，2003 年，第 5 页。

第三部分：学术影响

9 最初反响

要点 &—w

- 《物种起源》出版后就遭到批评，因为其物种变化理论反对生物是一成不变的主流观点，且《物种起源》没有提到所谓造物的神者。

- 科学界的主流承认物种的变化是存在的。达尔文在随后的《物种起源》版本中也做了一些改动，加入了一些涉及造物主的内容。

- 解剖学家托马斯·赫胥黎，和其他宗教界的朋友，如查尔斯·金斯利牧师*对《物种起源》进行了无畏的公开辩护。

批评

达尔文的著作《依据自然选择的物种起源》和他的进化理论引发了广泛的批评，主要集中在他忽略了智慧的造物主——上帝。物种变异这一概念也被认为是对神灵的亵渎，因为这暗示着上帝创造的生命是不完美的。不仅教会反对，达尔文的良师益友——地质学家亚当·塞奇威克也对他很是失望，对该理论所含的道德暗示深为恐惧。[1]

除此以外，在专业领域内，嫉贤妒能的风气也不可忽视。为了诋毁该理论，解剖学家理查德·欧文于1860年4月在《爱丁堡评论》中匿名撰文了一篇充满恶意的长文。欧文陈述达尔文提出的自然选择学说错误至极，主张"生物是在神的不断操作下完成的"。[2]

还有一些普遍的批评以科学为基础。达尔文很重视其中的一部分，经过认真地思考，在《物种起源》新版本里进行了反驳。第一个来自动物学家圣-乔治·杰克逊·米瓦特，他认为自然选择对眼睛等器官早期阶段的发育状况难以做出解释。[3] 他指出，如果眼睛是通过渐次累积性的方式产生出来，那就必定要经历一个阶段，期间它没有任何实际功能，这样一来，谈何选择性优势。

另一个反对意见于 1867 年来自苏格兰工程师弗莱明·詹金*。他认为，少数具有适应性特征的个体，在大量种群中与其他缺乏该能力的个体进行杂交时，自身的适应性特征很快就会消失。詹金认为在杂交过程中遗传因子会分化，在后代中优势特性也会被削弱。[4]

> "科学展示了过去不断的变化，并且在物质存在更广阔的发展中，模糊地指出了早期联系；但是，从神的意志创造自然和物质这个层面来讲，创世纪或起源，已超越物理哲学范畴。"
>
> —— 巴登·鲍威尔牧师：《创世纪哲学》

回应

达尔文在 1860 年出版的《物种起源》第二版（与第一版相隔仅两个月）中对宗教神学的批评进行了回应，增加了一些涉及造物主内容的句子。他还援引一位"著名宗教领袖"，即查尔斯·金斯利牧师的话来支持自己的自然选择观点——达尔文并未对此牧师指名道姓。他这样写道，牧师大人"相信，按照上帝的律例，真空的原始世界需要一种崭新的创造行为来填补，这种认识实在相当崇高，并且同样崇高的是，他逐渐认识到了神创造的一些最原始的生物有着自我发展进化的能力，除了能够满足形成需要的生物之外，还能生成新的生物。"[5]

对于米瓦特的批评，达尔文也做了回应，他辩称说器官在其发展的初期，就对生物体有益。例如，眼睛一开始只是对光敏感的器官，随着时间的推移，其他的功能也渐次出现，带来更多的好处，因此通过微小的进化步骤，眼睛这个器官才变成了现在复杂的样子。现在，人们都明白达尔文的解释是正确的："偶然出现的新事物可能有着微妙的优势。"然而，在他一生中，这个问题一直未能得到解决，这也是自然选择作为解释进化理论备受质疑的原因之一。[6]

达尔文也未能对遗传机制提供论据。事实上，在弗莱明提出这个问题之后不到一年的时间，遗传学*创始人格雷戈尔·孟德尔*就发表过一篇可以回答他的质疑的文章。不幸的是，孟德尔的作品直到 1900 年才被重新发现，得到人们的认可。孟德尔在 1866 年发表了一篇文章，展示了基因这种"看不见的因子"在编码过程中的作用。他展示了这些"因素"是不可再分的，因此詹金所设想的杂交过程中的混合是不可能的。[7]

出现了这些质疑，达尔文觉得有必要对自然选择学说进行一些补充，以此来说明进化过程也可以进行得很快。在《物种起源》第六版即最后一版中，他这样写道，自然选择"以一种很重要的方式，通过某些器官或部件的使用或放弃所产生的遗传效果，来发挥作用；以一种并不重要的方式，与适于生存的身体结构有关，无论过去还是现在，都随着外部环境状况的变化而直接产生作用。"[8]

这里，达尔文又回到法国生物学家让-巴蒂斯特·拉马克首先提出的获得性特征的遗传上来。达尔文还提出了一种新的理论——"泛生论"，根据这个理论，性细胞（精子和卵子）会吸收代表成年生物个体体内所有器官和组织的一整套粒子。这样，在成年阶段获得的特征就会传递给后代。这个理论与古希腊哲学家德谟克利特*所提出的理论相似。[9]

冲突与共识

尽管《物种起源》在说服科学界人士接受进化论方面取得了成功，但达尔文却没能让普通大众接受自然选择是物种变化主要机制这一观点，大部分原因是由于詹金的批评。1870 年以后，自然选择不再受到推崇，而拉马克的嬗变理论开始被广泛接受。这种观点认为，如果一个生物体为了生存能够根据环境的变化而适应性地改变身体结构，那么这些身体上的变化会遗传给它的后代。[10]

这一时期，还有其他一些理论应运而生，例如直生论*（现在看来则只是一个过时的生物假说而已）。西奥多·艾米尔*的著作《有机物发展原理：获得性特征遗传导致的有机进化》（1890），对直生论做了大力的宣传，表明生物体天生就具有一种倾向：以一贯稳定的方式不断进化，原因就是某种内在机制或所谓的推动力。[11]

1. R. 魏卡特：《从达尔文到希特勒：德国的进化伦理学、优生学和种族主义》，伦敦：帕尔格雷夫·麦克米伦出版社，2004 年。

2. 马克·雷德利：《如何阅读达尔文》，伦敦：格兰塔图书公司，2006 年，第 8—15 页。

3. 圣-乔治·杰克逊·米瓦特：《论物种的起源》，剑桥：剑桥大学出版社，2009 年。

4. 弗莱明·詹金："《物种起源》评论"，《英国北方评论》第 92 卷，1867 年 6 月第 46 期，第 277—318 页。

5. 查尔斯·达尔文：《论依据自然选择即在生存斗争中保持优良族的物种起源》第二版，伦敦：约翰·穆雷出版社，1860 年，第 481 页。

6. 马里奥·里维奥：《辉煌的失败：从达尔文到爱因斯坦：伟大科学家的异常错误改变了我们对生命和宇宙的理解》，纽约：西蒙与舒斯特出版社，2013 年。

7. 格雷戈尔·孟德尔："植物杂交试验"，《布鲁恩自然史协会》，1866 年。

8. 查尔斯·达尔文：《论依据自然选择即在生存斗争中保持优良族的物种起源》第六版，伦敦：约翰·穆雷图书出版社，1872 年。

9. 爱德华·J.拉尔森：《进化：科学理论的卓越历史》，纽约：现代图书馆出版社，2004 年。

10. 拉尔森：《进化》。

11. 西奥多·艾米尔：《有机物发展原理：获得性特征遗传导致的有机进化》，伦敦：麦克米伦出版社，1890 年。

10 后续争议

要点 ⚷━

- 《物种起源》意义深远，主要表现在我们理解人类行为——这种行为是固有的还是社会建构的，和人类"独特性"的概念上。

- 在受达尔文影响的生物人类学*中，人类被置于一个与之密切相关物种的动物环境中，人类行为和变异与动物的行为与变异一样受到研究。

- 达尔文的著作使学者们在人类和灵长类动物之间建立必然联系成为可能；人类现在被归类为动物，并归入人科或者类人猿科。

应用与问题

查尔斯·达尔文所著《论依据自然选择的物种起源》指出，物种通过自然选择而进化，但在 1870 年，达尔文对这种机制是否可以充分地解释进化问题却产生了严重的怀疑。但热烈支持这一理论的科学家——阿尔弗雷德·拉塞尔·华莱士和德国博物学家奥古斯特·魏斯曼*继续他的工作，发起了一个叫做"新达尔文主义"*的运动。

魏斯曼仔细地研究了一个概念，那就是父母亲能够将特定的特征（如钢琴技能或强壮的身体）遗传给他们的孩子，他甚至还做了一些实验来验证父母亲身上的伤疤是否会遗传给子代。在他 1883 年撰写的一篇论文中，他宣称获得性特征实际上无法遗传。性细胞（精子和卵子的合称）发育之初就被隔离开来，所以不管亲代身体发生什么样的变化，只要是隔离之后发生的，就不会受到影响。[1]

到 1885 年，他已经辨认出生殖细胞核就是遗传信息的载体。生殖细胞是生物体在有性繁殖时生成配子的生物细胞，由精子和卵子融合而形成。

1889 年，为了捍卫自然选择学说，华莱士撰写了《达尔文主义》。书中介绍了他自己关于物种形成的观点（物种是如何形成的过程），并强调了环境压力在迫使物种适应当地栖息地环境的重要性。种群在地域分布上如果比较分散孤立，配偶选择机会就少，就愈趋于分裂，最后就会形成两个独立的物种。[2] 这一理论被称为"华莱士效应"。[3] 当代研究证实了这一观点。[4, 5]

20 世纪初，植物学家雨果·德弗里斯*和卡尔·柯伦斯*重新发现了遗传学创始人格雷戈尔·孟德尔的工作。孟德尔的实验表明，后代会拥有父母亲本各自不同的特征，而非经混合之后所形成的特征。然而，科学家对孟德尔思想产生了误解，认为新特征、甚至是新物种都会突然出现。[6] 德弗里斯还进一步指出，新物种的产生并非通过自然选择，而是通过突变，即可遗传特征的突变。与达尔文渐变的观点相反，德弗里斯认为物种的进化是突然性的、急剧的。他的"突变理论"*建立在他对夜来香的观察研究之上。他观察到原植物有时会产生带有显著差异的后代，如在叶片形状或植株高度上的差异。德弗里斯就将这样的植物定义为新物种。

从本质上讲，德弗里斯思路正确，但理由错误：他观察到的大多数的变异是由染色体异常分离（成对的染色体进行分裂，并转移到细胞核的另一端）所致，而并非是突变。

在随后的几十年（1900—1930）间，出现了两种对立的思想学派。一派坚持达尔文自然选择的进化论观点，另一派则认为进化是一系列基因突变的结果，就像德弗里斯提出来的那样。为论证第二派理论，美国人类学家托马斯·亨特·摩根开始对普通的果蝇进行研究，并成功地将孟德尔对普通豌豆植物的研究和沃尔特·萨顿*的研究（确立了基因是遗传信息的载体）联系起来。摩根的实验表明，突变没有突然产生新的物种，但会在同一种群中增加新的变异。[7]他的研究详细地刊登在《孟德尔遗传学原理》中（1915）。[8]

从1930年到1950年，科学家们开始把孟德尔和达尔文的观点结合起来。达尔文的自然选择学说开始得到更广泛的认同，并且得到了来自遗传学和人口统计学这两大型领域新研究成果的支持。这个时期后来被称为新综合期，出现了进化生物学这一新的学科领域，依据进化理论，研究遗传机制。

> "我说，如果那样的话，要我回答这个问题：是愿意选择一个可怜的猿类，还是要一个天赋极高、能力非凡、又很有影响力的人做祖先，但此人用这些能力将嘲讽奚落带进庄严的科学讨论中——我会毫不犹豫地选择猿类。"
>
> —— 托马斯·亨利·赫胥黎："致达斯特博士的信"，
> 1860年9月9日

思想流派

新综合时期出现了现代进化论和社会生物学*两大思想流派，后者认为人类的行为都由进化产生。乌克兰进化生物学家狄奥多西·杜布赞斯基*是进化理论的杰出人物，他的《遗传学及物种起源》（1937）一书将达尔文的自然选择学说与孟德尔的遗传学和生物学学科的研究结合在了一起。[9]

在社会生物学中，社会行为被定义为能给个体产生有益结果的观念。英国科学家 W. D. 汉密尔顿*在 20 世纪 60 年代开展的亲缘选择*理论的研究帮助发展了这一新学科。[10] 汉密尔顿指出近亲之间彼此表现出更多的利他（无私）行为——这一现象被称为汉密尔顿法则*。汉密尔顿解释了利他主义，如昆虫中的真社会性*（即，存在着不育的昆虫群体）是如何从亲缘选择中发展而来的。[11]

当代研究

从 20 世纪 80 年代到 90 年代末，进化生物学的结构主义*观点得到了复兴；结构主义是智性思潮，在诸如人类学和语言学等领域有着极大影响力，这种观点（非常粗略地）认为某一语言或文化中不同的部分都是整个系统中的不可分的组成部分。结构主义的复兴，部分应该归因于生物学家布莱恩·古德温*和斯图尔特·考夫曼*所进行的研究工作，两人都强调了自我组织协调在进化过程中的贡献[12]（自我组织是指系统的组成部分相互作用，将本来无序混乱的系统变成秩序井然的系统）。

从 20 世纪 80 年代以来，新的数据不断积累，促使科学家认识到控制动物形成表型（可见的特性）的物质并不是那些形态各异的蛋白质集合。真正起作用的，是所有动物共有的一小类蛋白质在分布中的变化。[13] 这些蛋白质被称为**发育—遗传包**。[14] 这些知识对系统发育学（主要研究物种之间的进化关系的学科）、古生物学*（化石遗传学研究）和比较发育生物学*都产生了重大影响，并进而产生了一门全新的学科——进化发育生物学。[15]

现代生物学家不像达尔文那样关注自然选择能否解释物种的适应性，因为只有 5% 的遗传变化是适应性的。由于发现了脱氧核糖核酸（DNA*），（对此达尔文当然一无所知），现代科学家强调的是随机进化变化。(脱氧核糖核酸是基因携带的物质，为所有生物体的生长和正常运行提供指导。)

现在普遍认为两个主要过程引起了进化：自然选择和随机遗传漂变（这一过程指的是遗传信息在动物繁殖期内的变化）。正如达尔文所说，进化不仅仅是由自然选择所驱动。如果一个基因或等位基因 *（基因变体）有两个同样好的版本，并且一个比另一个更幸运，在一个种群中加以传播，经过数代时间的衍变，进化可以偶然发生。

举一个简单的例子，一个孩子从母亲那里继承了黑发的基因代码，从父亲那里继承了红发的第二基因代码。后来，这个孩子又有了自己的孩子。只能传递一个基因的副本，偶然将黑发代码的基因传给孩子。经过数代，情况再次发生，黑头发就会在这一人群中占主导地位，而红头发的代码则会消失。然而，达尔文仅了解生物体的可观察形式，因此更多的关注这些外在特征的演变。[16]

---

1. 弗里德里希·利奥波德·奥古斯特·魏斯曼：《水螅水母性细胞的出现：建构该种群结构与生命体征》，耶拿：菲舍尔出版社，1883 年。

2. 阿尔弗雷德·拉塞尔·华莱士：《达尔文主义：自然选择学说阐述及其应用》，伦敦：麦克米伦出版社，1889 年。

3. 爱德华·J. 拉尔森：《进化：科学理论的卓越历史》，纽约：现代图书馆出版社，2004 年。

4. 米歇尔·杜林克斯和汤姆·J. M. 范多恩："交配选择和优势改良：消除杂合子缺陷的替代方法"，《进化》第 63 卷，2009 年第 2 期，第 334—352 页。

5. J. 奥勒顿："物种形成：开花时间与华莱士效应"，《遗传》第 95 卷，2005 年第

3 期，第 181—182 页。

6. 拉尔森：《进化》。

7. 拉尔森：《进化》。

8. 托马斯·亨特·摩根等：《孟德尔遗传机制》，纽约：亨利·霍尔特出版社，1915 年。

9. 拉尔森：《进化》。

10. 乔尔·L.萨克斯："物种内部与物种之间的合作"，《进化生物学杂志》第 19 卷，2006 年第 5 期，第 1415—1418 页。

11. 马丁·A.诺瓦克："合作演化的五项原理"，《科学》第 314 卷，2006 年，第 1560—1563 页。

12. 彼得·康宁：《整体达尔文主义：协同控制论以及进化的生物经济学》，芝加哥：芝加哥大学出版社，2010 年，第 95—99 页。

13. 约翰·R.特鲁和西恩·B.卡罗尔："生理和形态进化中的基因协同合作"，《细胞和发育生物学年度回顾》第 18 卷，2002 年，第 53—80 页。

14. 克里斯蒂安·卡内斯特罗，横井喜树和约翰·H.波斯特尔维特："进化发育生物学和基因组学"，《自然评论：遗传学》第 8 卷，2007 年第 12 期，第 932—942 页。

15. 贾梅·巴古纳和詹迪·加西亚-费尔南德兹："进化发育生物学：漫长而曲折的道路"，《国际发展生物学杂志》第 47 卷，2003 年第 7—8 期，第 705—713 页。

16. 马克·雷德利：《如何阅读达尔文》，伦敦：格兰塔图书公司，2006 年。

11 当代印迹

要点 🔑

- 《物种起源》自出版至今都十分有价值。

- 关于固有行为（达尔文称之为"本能"）与社会建构主义*（认为人的行为通常都是由周围的文化环境构建而成的观点）的争论十分激烈，争论同样激烈的还有人在动物界里的独特性问题。

- 激烈反对进化论的一些宗教团体仍然存在，他们担心那样会导致无神论。

地位

　　受查尔斯·达尔文在《论依据自然选择的物种起源》一书中提出的自然选择理论的启发，产生了一个叫做社会生物学的新学科，该学科尝试着将进化论的理论框架运用于人类社会的研究。

　　有人认为，人类的某些行为在出生之际就在我们的基因里进行了"编程"，这种观点被称为生物决定论*。社会生物学理论研究一直由理查德·道金斯*和爱德华·奥斯本·威尔逊*等生物学家所引领。在《自私的基因》（1976）一书中，道金斯指出，不管我们怎么努力，都无法逃脱基因的支配作用。[1] 威尔逊将这一思想加以拓展，广泛应用于社会研究中，他说，所有社会"无论多么平等，总会因为男性和女性之间固有的基因差异，使得男性拥有比女性更多的权力"。[2]

　　而史蒂芬·杰·古尔德和理查德·莱万廷*等进化生物学家对这一方法进行了批评，认为那是对人类行为的还原论*（是过于简化的一种阐释）。最近的社会科学研究表明，决定性别差异的因素主要是趋势，而非某种确定的行为。[3,4] 另一方面，强硬决定论者，例如进化心理学家史蒂文·平克*将行为定义为"男性"或"女性"。在 1982 年的"泰纳讲演"中，莱万廷教授提出的观点让我们洞察了为什么决定论思想盛行的原因："我们的个体、种族、国家和性别之间众多的不平等所产生的问题就是不平等的现实与我们的社会理应遵循的平等意识形态之间的明显矛盾。"[5]

因此，科学已经取代宗教成为知识领域的终极权威。过去，社会一切的不平等被认为是上帝的旨意。而现在，一些科学家正在尝试着用生物学中的观点来证明社会不平等也是正当合理的，他们声称男女之间不平等是由于两者大脑生物构造内在的差异所造成。这些科学家因用科学的观点回答个人、社会或政治性质的问题强化主流意识形态而遭受批评。

> "你都可以给亚里士多德上一堂课，并且你的知识会让他发自肺腑地激动。然而，你不仅可以比他对这个世界了解得更多，还可以更深刻地理解万事万物的运行原理。这就是有大师的好处，如牛顿、达尔文、爱因斯坦、普朗克以及他们的同事们。这里我说的不是你比亚里士多德更聪明，或者更睿智。就我所知，亚里士多德算得上有史以来是最聪明的人了。但这不是问题的关键。问题的关键是科学研究探索是建立在前人累积的基础上的，而我们属于踩在巨人肩膀上的后来者。"
>
> —— 理查德·道金斯："科学、虚妄和对奇观的嗜好"

互动

自然选择表明人类是漫长进化过程的产物，因为在众多动物中，人是专业化的动物。这种观点在 19 世纪具有争议性，因为它挑战了圣经的例外主义学说，这一学说宣称上帝创造的顶峰便是人类。这个极具争议的观点在维多利亚时代造成了人类身份的危机，因为许多人拒绝相信自己与动物有着如此紧密的联系。[6]

即使在今天，人类独特性这一问题与 1859 年《物种起源》出版时一样，仍受到人们广泛的争论。[7] 人类学家金·希尔*认为，把我们从其他动物区别开来的是我们对文化与合作的依赖。进化心理学家约瑟夫·考尔*和迈克尔·托马塞洛*指出，大猩猩能理解意会同伴的意图，而在独一无二的人类语境下，是可以分享意图的。[8] 托马塞洛得出结论：这就证明我们人类和其他类人猿之间存在有认知鸿沟。[9]

然而，"人与高等动物之间在心智上的差异虽然巨大，但只是程度不同而已，并非本质性差异。"[10] 毫无疑问，这种认识仍然是反对人类例外论的主要观点。马克·哈里斯等神学家认为，用于证明人类独特性的科学证据，仅仅是人类与其他动物之间的定性差异，因为没有特征将人类与其他灵长类动物区别开来。[11] 对人类起源的研究已经揭示出我们人类与生物表兄弟之间的相似生物属性多于不同之处。

持续争议

相反，近百年来，进化论的观点在主流科学界没有什么争议。20 世纪 30、40 年代以来，当现代进化理论将遗传学纳入了达尔文的理论体系后，否认进化论思想的人大多数都来自信奉上帝的创造神话的宗教团体，反对进化论的论点包括对证据、科学论证方法、道德和合理性的异议。[12]

智慧设计运动中一个核心文本是由律师菲利普·约翰逊撰写的《审判达尔文》（1991）一书。书中使用了法律框架组织论证。举例来说，法律术语"毋庸置疑"（一般用于极罕见情况下，表示有足够证据确定某事确凿无疑）被用来摧毁科学理论。[13] 然而，进化生物学家史蒂芬·杰·古尔德指出：像这样的法律论点用在科学的探讨上本身就是不妥当的，因为"科学并不是声称建立确定性的行为准则"。[14]

当然，进化论促进了无神论这样的观点也使进化论遭到强烈反对。[15] 神创论者声称，进化论的支持者都是无神论者，他们看不到物质原因和事实的背后，因此对存在本质是不可能做到全面理解的。[16] 但这些说法是站不住脚的；在 2014 年进行的一项民意调查显示，美国有 40% 的科学家是相信神的存在的，这与在美国公众中所作的调查结果相似。[17,18]

宗教文学家坚定不移地支持创世神话，但有证据表明，创世纪*的圣经书是对早期神话故事的复述——其中包含古代中东吉尔伽美什神话故事的主题和古代美索不达米亚人对世界的理解。一个美索不达米亚神话讲述了宁悌女神被派去医治另一位叫恩基的神，后者正遭受着疾病的摧残。苏美尔语中"宁悌"一词两义（同时可指"肋骨"与"生命"），这在创世纪里是一个反复出现的主题：夏娃由亚当的肋骨创造而成。[19]

事实上，基督教宗教学者布鲁斯·沃尔特克和康拉德·海尔斯就告诫人们不要对创世神话只作字面解读，因为它与美索不达米亚科学*和宗教原则一致，那样只能导致进化受阻。在他们看来，当代科学知识应该被纳入创世故事建构之中。[20] 罗马天主教教派正是这么做的，他们通过主张有神进化论*（此观念认为进化是付诸行动的过程，由上帝之手进行引导），进而实现了信仰神和进化两者和谐一致。

1. 理查德·道金斯：《自私的基因》，牛津：牛津大学出版社，1990 年。

2. E. O. 威尔逊："人间体面有动物的特征"，《纽约时报》，1975 年 10 月 12 日。

3. B. J. 卡罗瑟斯和 H. T. 赖斯："男人和女人来自地球：审视性别的潜在结构"，《人格与社会心理学》第 104 卷，2013 年第 2 期，第 385—407 页。

4. 达普娜·乔伊等："生殖器以外的性：人脑万花筒"，《美国国家科学院院刊》第 112 卷，2015 年第 50 期，第 15468—15473 页。

5. R. C. 莱万廷："生物决定论"，《关于人类价值观的泰纳讲演》，犹他州立大学，1982 年 3 月 31 日和 4 月 1 日，登录时间 2016 年 2 月 5 日，http://tannerlectures. utah.edu/-documents/a-to-z/l/lewontin83.pdf。

6. 爱德华·J. 拉尔森：《进化：科学理论的卓越历史》，纽约：现代图书馆出版社，2004 年。

7. H. 古德伯格："重申人类的独特性"，《今日心理学》，2010 年 11 月 8 日。

8. J. 布劳尔，J. 科尔和 M. 托马塞洛："黑猩猩真的知道同伴在竞争的状态下会看到什么"，《动物认知》第 10 卷，2007 年，第 439—448 页。

9. M. 托马塞洛等："意图的理解与分享：文化认知探源"，《行为与脑科学》第 28 卷，2005 年第 5 期，第 675—735 页。

10. 查尔斯·达尔文：《人类的由来及性选择》，伦敦：约翰·穆雷出版社，1871 年，第 82 页。

11. 马克·哈里斯："人类独特性：人类是进化的巅峰吗？"，《科学与宗教 @ 爱丁堡》，2014 年 9 月 7 日，登录时间 2016 年 2 月 16 日，http://www.blogs.hss.ed.ac. uk/science-and-religion/2014/09/07/human-uniqueness-and-are-humans-the-pinnacle-of-evolution/。

12. 彼得·库克：《进化与智慧设计：小题大做？双方的论点》，澳大利亚：新荷兰出版社，2007 年。

13. 菲利普·E.约翰逊：《审判达尔文》，伊利诺伊州唐纳格罗夫：通视出版社，1991 年。

14. S.J.古尔德："弹劾一名自封的法官"，《科学美国人》第 267 卷，1992 年第 1 期，第 118—121 页。

15. 李·斯特罗布：《神创者案例：记者调查指向上帝的科学证据》，密歇根州大急流城：崇德凡图书，2004 年。

16. 菲利普·E.约翰逊："达尔文教堂"，《华尔街日报》，1999 年 8 月 16 日。

17. 拉里·威萨姆："许多科学家在进化中看到了上帝之手"，《国家科学教育中心的报告》第 17 卷，1997 年 11 月至 1997 年 12 月第 6 期，第 33 页。

18. 布鲁斯·A.罗宾逊："美国公众对进化和神创的信念"，登录时间 2015 年 4 月 14 日，www.ReligiousTolerance.Org。

19. 塞缪尔·亨利·胡克：《中东神话》，多佛出版社，2013 年，第 115 页。

20. 康拉德·海尔斯，《创造的意义：创世纪与现代科学》，路易斯维尔：威斯敏斯特·约翰·诺克斯出版社，1984 年。

12 未来展望

要点 🔑

- 《物种起源》提到的理论从未被证明有错误，在未来很长时间内对科学研究有重要影响。
- 今天，DNA 和遗传学研究的兴起都有赖于自然选择学说理论，并且未来很可能会继续如此。
- 《物种起源》所提出的理论构成了当代所有生物科学的基础。

潜力

查尔斯·达尔文所著的《论依据自然选择的物种起源》讨论了过去地球上生命的进化过程，但很多人想知道人类进化的未来。

威尔士遗传学家史蒂夫·琼斯*认为，人类进化正在放缓，因为我们不再受自然选择的左右——适者生存的个体不再推动进化过程。达尔文在 19 世纪发表《物种起源》时，只有不到一半的英国儿童能够活到 35 岁。现在这个数字已经达到将近 95%，[1] 主要是因为医学在治疗各种疾病方面取得了很大的进展。现在，即便是最弱的个体也能继续存活并养育子女。用琼斯的话说，"达尔文的进化机器已经失去了往日的力量。"[2]

进化心理学家杰弗里·米勒*却有不同的看法。由于出现了诸如飞机等现代交通运输方式，病毒和细菌病原体现在可以更轻易地散播到全球各地。米勒因此预测：流行性疾病对于塑造人类免疫系统将变得很重要，并将使人类后代拥有更强大的免疫系统。

尽管《物种起源》在许多方面已失去现实意义，但它提出的观点仍占主导地位，并超越了科学的范畴，影响着我们今天绝大多数人的生活。

> "通过对血红蛋白分子和其他分子的氨基酸序列具体而又详细的测定，有可能获取进化过程中的众多信息，从而阐明物种起源问题。"
>
> —— 莱纳斯·鲍林:《分子病与进化》

未来方向

古希腊哲学家亚里士多德曾将生命设想成为一个存在巨链，每一个生物体在这个巨链的等级中占据着自己的位置，而且在"进化论话语体系中仍然有一种倾向，将生命历史描述为向更加复杂状态进展的过程。"[3] 但是，也会出现这些情况：比如无胆绦虫或盲洞穴鱼这类简单生物体，是从复杂生物发展而来的。这种情况被称为"还原进化"。

研究还发现，生物体偶尔会丧失曾被认为是生存的必备技能，而不会影响任何生存或繁殖能力。这种情况是在海洋蜉蝣生物——原绿球藻中首次被发现的，这是一种较为常见的光合微生物（一种可将太阳光转化为能量的生物体）。研究人员发现这种生物已经失去了有助于中和过氧化氢（一种可以破坏细胞的化合物）的基因。[4]相反，它依靠邻近的微生物为它消除环境中的过氧化氢。

对于微生物来说，携带基因和制造蛋白质都需要消耗大量能量，所以抛弃某些基因使它们能更加高效地存活。首先观察到这一现象的研究人员给它命名为"黑皇后假说"*——这一原则的含义就是：如果微生物邻近有其他物种可代替行使某些重要功能，自然选择就会驱使微生物放弃相关功能。[5] 进化会让发挥作用的施助生物与受助生物和平共处。施助生物的适应能力没有降低，反而因为形成的共存关系向着共生性 * 的方向进化，两种完全不同的生物体由于这样的安排能够最终相互合作，共同受益。因此，这些施助者并不是社会利他主义者，不过是输了比赛而无法遗传自身基因的群体，困在功能执行者这一角色中。[6]

达尔文强调竞争和冲突，而加拿大生物学家布莱恩·古德温则认为生存问题的本质其实就是寻找一个适合自己的小环境。古德温还认为，幸存下来的生物体也并不见得比那些已经灭绝的生物多高级。相反，进化就像与生物在"共舞"，目的"只是寻求一块可供生存下去的空间"而已。[7]

小结

进化思想虽在达尔文之前便已出现，但达尔文的自然选择理念让进化论看起来更趋合理：个体存在差异，随机选择使一些个体比其他个体更能适应自身环境，适应性更好的个体就会大量产生后代。丰富多彩的动物物种的产生不是通过上帝的创造，而是通过进化完成的。最终，达尔文将科学和宗教中分离开来。通过摈弃造物主，科学地对自然现象进行解释变得至为重要。[8] 不管是过去还是现在，达尔文因为这一创新，将一直受到世人的称颂。

达尔文创立了进化生物学这门学科，自然选择为其基本原则。他在这个领域的另一个贡献则是主张物种是在一个漫长的过程中发生变化的，这就是所谓的渐进性过程。他还用分支方式而非线性方式来绘制进化蓝图，暗示所有物种都来自一个唯一且独一无二的源头。达尔文认为选择发生在个体层级，现在我们知道这发生在基因层面上。近年来因此还兴生了一个叫做基因疗法*的医学新领域，使得治疗遗传疾病成为可能，如囊性纤维化*、癌症以及其他一些传染疾病（包括艾滋病病毒 HIV*）。医生可以在患者子宫内（宫内）进行这种治疗，甚至可以在产前进行治疗某些性命攸关的疾病。尽管基因疗法可以使人类在未来免受特定遗传疾病的困扰，但也可能影响胚胎的发育，对此我们无法作出预测，或者会产生长期的副作用，[9]也未可知。因此这仍是一个带有争议性的问题。

---

1. 英国国家统计局："2010 年英国死亡率统计"，登录时间 2016 年 2 月 5 日，http://www.ons.gov.uk/ons/rel/mortality-ageing/death-in-the-united-kingdom/mortaliy-in-the-united-kingdom-2010/mortality-in-the-uk-2010.html。

2. 史蒂夫·琼斯：在剑桥大学举行的纪念达尔文 200 周年纪念和《物种起源》150 周年的演讲。

3. 杰弗里·J.莫里斯，理查德·林斯基和埃里克·辛塞尔："黑皇后假说：适应性基因缺失导致的进化依存"，《兆比》第 3 卷，2012 年第 2 期，第 1—7 页。

4. 莫里斯："黑皇后假说"，第 1—7 页。

5. 莫里斯："黑皇后假说"，第 1—7 页。

6. 莫里斯："黑皇后假说"，第 1—7 页。

7. 布莱恩·古德温：《豹斑是如何改变的：复杂性的演变》，新泽西州普林斯顿：普林斯顿大学出版社，2001 年，第 98 页。

8. 恩斯特·迈尔："达尔文对现代思想界的影响"，《美国哲学学会会刊》第 139 卷，1995 年 12 月第 4 期，第 317—325 页。

9. 如有意了解关于基因疗法应用的相关问题的综述，请参阅索尼娅·Y.亨特："关于治疗方法上的争议：基因疗法、人工授精、干细胞和药物基因组学"，《自然教育》第 1 卷，2008 年第 1 期，第 222 页。

术语表

1. **废奴主义**：在西欧和美洲结束奴隶制的运动。

2. **等位基因**：基因的一种变体形式。一些基因可以以各种不同的形式出现。

3. **利他主义**：关心他人的福利并且为此甘心奉献。利他主义行为指的是毫不利己、专门利人的行为。

4. **解剖学家**：专门从事动物解剖的人。

5. **人类学**：研究人类文化和社会生活的学科。

6. **节肢动物**：该名称指代无脊椎动物，身体分节，有连接的副肢。昆虫、千足虫、甲壳类动物和蜘蛛都属节肢动物。

7. **无神论**：否定神明存在的信仰。

8. **"贝格尔"号**：HMS（"女王陛下之船"）贝格尔号是一艘皇家海军舰艇，查尔斯·达尔文曾搭该船进行科学航游，用了五年时间完成世界航行，收集标本并创立了自然选择学说。

9. **行为生态学**：利用进化理论和动物生活环境（包括捕食者和天气等因素）来理解行为的一个科学领域。

10. **生物人类学**：又称"物理人类学"，是把人类视为动物所进行的生物研究。

11. **生物决定论**：这种观点认为所有（或大多数）人类行为都是由基因所决定的，而非教养或个人选择。

12. **生物遗传**：能够从亲代传给子代的各种遗传特征。

13. **黑皇后假说**：这种观念认为微生物，如果在自身环境中的其他微生物可以为其执行功能，有时它们就会丧失该执行能力，即使该功能对它们的生存至关重要也不例外。这种适应促进微生物生活在一个相互合作的环境当中。

14. **资本主义**：一种社会经济体系，其中贸易和工业掌握在个人手中，并为获取个人的利益而展开。

15. **比较解剖学**：对不同物种的解剖结构进行比较和对比的研究。

16. **比较发育生物学**：利用生物体的自然变异和差异来了解各级生命形式增长模式的一种学科。

17. **神创论**：认为自然世界及其生物体都是由一个神或多个神所创造的一种观念。

18. **囊性纤维化**：一种可危及生物体重要器官功能的遗传性疾病。

19. **变异性遗传**：该术语描述的是随着时间和世代的演进，赋予生殖优势的某些特征是如何在某一种群中变得更加普遍的。

20. **趋异**：一个物种分裂成两个或两个以上的独立物种的现象。

21. **DNA**：脱氧核糖核酸的缩写，是一种携带着绝大多数基因指令的分子，在所有生物体和许多病毒的生长发育过程和繁殖过程中发挥作用。

22. **生态区位**：生物体在社区中的生态角色。此概念尤其适用于食物消费。

23. **启蒙运动**：17世纪和18世纪在西欧出现的思想文化运动。比起传统思想及神干涉人类事务的观点，启蒙运动强调的是理性。其时著名的学者有法国作家伏尔泰和卢梭，以及德国哲学家伊曼努尔·康德。

24. **侵蚀**：由风、太阳或水驱动的地质过程，会破坏土壤或岩层结构。

25. **本质主义**：此观点认为对于任何特定实体（例如动物或物理对象）来说，都存在着一组必要属性来反映其身份和作用。这个概念源于希腊哲学家柏拉图和亚里士多德的著作。

26. **真社会性**：这一概念描述生活在合作性群体中的生物，其中专门从事繁殖活动的通常只有一个雌体和数个雄体，其余个体全部都与繁殖无关，只是照料年轻的成员、或为他们提供保护，或为整个群体提供保障服务，比如白蚁、蚂蚁和裸鼹鼠。

27. **进化**：生物体遗传特征中出现的代与代之间的变化。

28. **进化生物学**：对物种随时间而变化进行的研究。该学科的子领域包括分类学、生态学、种群遗传学和古生物学。

29. **进化心理学**：进化语境下的心理学研究。

30. **进化论**：社会始于"原始"并趋向"文明"的观念。

31. **灭绝**：生物或物种最终消亡的生物学术语。

32. **基因**：能够对功能性蛋白质产品进行编码的一个 DNA 片段。它是遗传的分子单位。

33. **基因疗法**：通过改变或替代患者细胞中的基因来治疗由遗传异常或缺陷引起的医学疾病。

34. **创世纪**：摩西五经中的第一本书（《创世纪》、《出埃及记》、《利未记》、《民数记》、《申命记》），是犹太教和基督教经文的一部分。这本书描述了地球和人类的创造、人类种族的扩张以及亚伯拉罕及其后代的故事。

35. **遗传漂变**：由纯粹随机抽样引发的基因变异所导致的非适应性变化。

36. **遗传学**：对生命形式中的基因和遗传变异的研究。

37. **地理学**：研究行星表层的土地和环境以及人类与其生态区位和环境相互作用方式的学科。

38. **地质学**：研究物理地球或任何天体，包括多种板块构造或演化等领域。

39. **渐进性**：在进化研究中，物种在中间阶段随着时间而世代变化的概念。

40. **存在巨链**：古希腊哲学家柏拉图首先提出的一个概念。主要思想是生物都属于一个层次结构的某个地方。所以每样事物都有自己的位置，因为这就是上帝所希望的。

41. **1832 年改革法案**：1832 年在英格兰和威尔士颁布的法案，赋予以前被剥夺公民权的大批男性公民投票权。苏格兰和爱尔兰在同一年颁布了类似的法案。

42. **群体选择**：个体为集体或整个物种的利益而进行选择的过程，是由英国动物学家怀恩－爱德华兹于 1986 年提出的理论。但自那时以来一直受到怀疑。这个概念第一次得到支持是来自诺贝尔奖得主——动物学家康拉德·洛伦茨等科学家。

43. **汉密尔顿法则**：是对利他主义演变所需的条件的解释，例如，一个生物帮助自己的近亲以自己的幸福为代价生存下去。公式是 r×B>C，其中 r 代表的是个体间的亲缘关系程度，B 是对受益人的好处，C 是执行该特定行为的成本。

44. **HIV**：人类免疫缺陷病毒。这是一种逆转录病毒。受感染个体的免疫系统功能降低。艾滋病毒是艾滋病的一个病因。

45. **杂交**：两种不同物种、亚种、偶尔甚至不同属的后代。

46. **研究先驱**：该术语指的是在某领域内一些开创性的成就，被公认是同类研究中的奠基性成就，标志着一个重要的发展阶段。

47. **智慧设计**：描述地球是由上帝设计和创造的术语。

48. **异性选择**：由于各具特殊性，异性间所出现的相互吸引的现象及过程。

49. **同性选择**：同性成员相互竞争寻找交配机会的现象及过程。

50. **亲缘选择**：在具有亲缘关系的个体之间经历进化时期所表现出来的、受到偏爱的行为。

51. **拉马克主义**：在达尔文时代，该术语指的是物种的嬗变（改变），这违背了当时正统的科学认识：即只存在固定状态的生物。近年来通常表示一种错误观念，即亲代在一生中所形成的某些特质会遗传给后代。

52. **自由主义**：政府应该负责保护社会中每个人的自由和平等的观念。

53. **生命科学**：论述生物体的科学领域，例如神经科学、植物学和病毒学。

54. **语言学**：研究语言结构和性质的学科。

55. **林奈学会**：一个在伦敦建立的旨在促进自然历史和分类学（对所有生物的分类）研究的学会。

56. **马尔萨斯灾难**：一种假设的场景，一个社会的所有成员都被迫退回到自给自足的农耕经济以养活家庭。预计这种情况将会在人口增长速度超过农业生产的速度时发生。

57. **哺乳动物**：属于脊椎动物，该种动物的幼崽通过吸食母体乳腺中奶水长大。这类动物典型例子有海豚、人类、狗和奶牛等等。

58. **唯物主义者**：指为所发生的物质现象单纯从物质的角度探寻原因的人。

59. **美索不达米亚科学**：对古代美索不达米亚地区的自然和文化现象，包括真实的或是想象的，进行的各种形式的学术研究。该研究主要包括苏美尔人在书写方面的发明创造、对时间的划分（比如 60 秒钟和 60 分钟为单位的时间）以及车轮的发明。

60. **现代进化生物学 / 现代进化论 / 新综合论**：这是发生在 20 世纪 30 年代到 40 年代间对达尔文进化理论的改进，融入了遗传学和其他新理论，如亲属选择和人口统计学等。

61. **怪物**：达尔文称突变为"怪物"。在开创性的遗传学家格雷戈尔·孟德尔发现基因之前，人们还不能清楚地理解遗传学和遗传特征。

62. **突变**：生物体基因，即其遗传物质组序列的变化。

63. **突变理论**：达尔文指出，新物种的出现不是由持续变异、而是由突然变异形成的，称为突变。突变理论于 1901 年由生物学家雨果·德弗里斯首次提出，指出突变是跨越连续几代后得以遗传的。

64. **共生性**：分属不同物种的两种生物体彼此合作，共同受益。

65. **自然选择**：解释物种变化的一种机制。查尔斯·达尔文和博物学家阿尔弗雷德·拉塞尔·华莱士指出，生物体在不同的环境下努力生存；生物将其最适于生存的特性遗传给后代，而其余存活时间较短、无法繁殖的物种则会灭绝。

66. **博物学家**：研究自然世界的学者。

67. **自然主义谬论**：一种认为东西只要是"自然的"，就是"好的"观念。

68. **新达尔文主义**：具体指科学家，即华莱士和魏斯曼，抛弃了拉马克式的获得性遗传理论，大力提倡自然选择作为物种进化的根本原因的这一段时期。这个术语是由乔治·罗曼斯在 1895 年创立的。

69. **虚无主义**：这种哲学观点认为，人生最终没有意义，人们的所作所为也毫无意义。坚持这种哲学立场的人同样否认人体和灵魂的二元性。

70. **非国教派**：也称为一神论派，这是一个 17 世纪时的用语，指的是当时信仰英格兰新教的教徒（特别是清教徒和卫理公会教徒），因为新教非官方教会，故而得名。

71. **海洋学**：对海洋的科学研究。

72. **直生论**：生物学家西奥多·艾米尔所创的理论，认为物种的变化是生物个体内部力量的直接结果，目的是改善当前的物种类型。

73. **古生物学**：对古代存在的、现在保存在岩石中的动植物所进行的科学研究。该领域的科学家与地质学家紧密合作，以确定岩石和生物体化石遗迹距今存在的时间。

74. **泛生论**：达尔文所创立的关于遗传的理论，但自创立以来，就被证明是错误的。根据此理论，来自生物体内每个细胞的信息（以粒子的形式）都会传递给生殖器官；在生殖器官中，所有此类信息都进行合并，进而形成精子和卵子。

75. **范式**：一种被普遍接受的思维方式或行为方式。

76. **亲代物种**：能够产生其他物种的物种，通常需要漫长的时间。通常指两个不同物种间最近的共同祖先。

77. **表型**：个体的可见性特征。个体的表型不仅受其基因（或基因组成）的影响，而且受其成长和生活环境的影响。

78. **布里尼学会**：由爱丁堡大学的学生组织的私人会员俱乐部，学会活动围绕科学家所发表的自然历史相关问题著述展开阅读和讨论。

79. **实证主义**：知识只能由对经科学所证实的事实进行分析的方式进行推进。

80. **灵长类动物学**：对灵长类动物进行的研究，人类和类人猿均属灵长类。

81. **还原论**：将相互关联的多种不同的哲学立场或理论化简为"更简单"或更"基础"的一种哲学思想。

82. **相对主义**：该哲学概念认为不存在绝对真理，真理都是相对的。通常应用于社会文化人类学语境，表明人类行为和信仰应该在特定文化背景下加以理解。

83. **科学优先**：一个通用术语，指在特定领域具有开创性并被视为重大进展的工作，如 DNA 双螺旋结构的发现。

84. **社会建构主义**：认为行为方式往往是由周围文化力量所建构起来的一种理论。

85. **社会生物学**：考察人类社会行为进化的源头和功能的科学领域。

86. **社会文化人类学**：对目前或某一特定时期人类文化变异性所进行的研究。

87. **性别选择**：该理论认为雄体为获取与雌体的交配权而相互竞争，而雌性（较少指雄性）也会对与想要交配的对象进行选择。因此，一些被认为"有吸引力"的特性，如孔雀尾巴，可能就会比其他特性更多地遗传给后代。

88. **社会达尔文主义**：将达尔文的自然选择机制尝试性地应用于人类社会结构的研究。这意味着某些种族比其他种族得到更多的"进化"，而"成功"的社会也是最适合人类的，对力量强大或最具竞争力的个体会有更大的回报。

89. **物种形成**：一个新物种形成的过程。当一个物种分裂成两个或两个以上的独立世系时就会形成新的物种。

90. **物种**：生物中最广大的分类单位，其中个体能够通过两性繁殖，生育的后代可以继续繁衍新的后代。

91. **结构主义**：理解人类文化的一个理论框架，描绘了人际互动的模式和相互关系，随之产生的模式或结构用来更好地理解人类文化。

92. **亚种**：一群可以杂交的生物，有时在野外进行自然杂交。在分类学中，物种列在亚种之后。这是新物种形成过程中经历的第一阶段。

93. **适者生存**：这个术语指的是最适合生存和生育后代的个体。因此，在生物学中，"适者"一词指的是那些能够在一生中生产最多后代的个体。

94. **分类学**：指在生物学中，对现存生物的系统分类。1735年，瑞典博物学家卡尔·林奈对分类学和命名法（对生物体的科学命名）进行了完全的革新。

95. **禁酒运动**：一系列大规模的社会和政治运动，旨在取缔饮用或减少用酒精的合法性，这项运动于18世纪中叶至20世纪初在欧洲和北美的政治影响力达到顶峰。

96. **有神进化论**：这种思想认为进化过程得以进行全部有赖于上帝之手的指导。

97. **神学**：对宗教思想进行的系统研究，通常以梳理宗教经文的方式进行。

98. **嬗变（物种的）**：法国生物学家让-巴蒂斯特·拉马克于1809年发明了这一名称，用来描述物种从单一的原型到多种多样形态变化的可能性。

99. **均变论**：这种理论认为，当今改变地球的力量在过去同样起着作用；这一理论由苏格兰地质学家詹姆斯·哈顿于1788年首次提出。

100. **动物学**：对动物的研究。

人名表

1. 查尔斯·贝尔爵士（1774—1842），苏格兰解剖学家，因发现感觉神经和运动神经的作用而闻名。他所著作品还包括关于宗教、哲学和神学的著述。

2. 珍妮特·布朗（1950年生），英国科学史学家，也是查尔斯·达尔文传记作家中最著名的一位。以《查尔斯·达尔文传（第一卷）：航游》和《查尔斯·达尔文传（第二卷）：地方的力量》而闻名。

3. 乔治-路易斯·勒克莱尔，孔蒂德·布封伯爵，（1707—1788），法国博物学家、数学家，还曾任著名的法国植物园总监一职。在启蒙时期乔治就被誉为自然史之父。他最著名的作品当属《自然史》（1749—1788），是一部36卷本的百科全书，描绘了动物及矿物王国的全貌。

4. 约瑟夫·考尔（1967年生），西班牙比较心理学家。他的研究重点是人类以外灵长类动物认知及其与人类智能的比较。如今他担任德国莱比锡马克斯·普朗克研究所的沃尔夫冈科勒灵长类动物研究中心任主任。

5. 罗伯特·钱伯斯（1802—1871），苏格兰科学家和出版商。他于1844年以匿名方式撰写了一篇主要讨论在达尔文之前的进化论思想探源的文章，收入到了一部名为《自然创造史的痕迹》的著作中。

6. 卡尔·埃里希·柯伦斯（1864—1933），德国遗传学家和植物学家。他著名的发现是生物遗传原理，以及对格里戈尔·孟德尔在豌豆植物和遗传方面研究的重新发掘。他与雨果·德弗里斯可谓是同时但又是独立完成了这一任务。

7. 乔治·居维叶（1769—1832），法国博物学家和动物学家，以宗教和科学理由支持物种的固定不变性。他因著有《动物自然历史元素表》（1798）而闻名。

8. 艾玛·达尔文（1808—1896），祖姓韦奇伍德，查尔斯·达尔文的妻子，和他共同养育了 10 个孩子。她是达尔文的大表妹，都属于韦奇伍德这一名门望族（韦奇伍德陶器制造商）。

9. **伊拉斯谟斯·达尔文**（1731—1802），查尔斯·达尔文的祖父。他是一位杰出的科学家和医生，提出了关于物种嬗变的想法。

10. **罗伯特·沃林·达尔文**（1766—1848），英国医学博士、博物学家查尔斯·达尔文的父亲。

11. **理查德·道金斯**（1941 年生），科学作家、进化生物学家和杰出的无神论者。他以 1976 年出版了《自私的基因》一书而闻名，该书论证了自然选择发生在基因层面。

12. **德谟克利特**（公元前 460—370），古希腊前苏格拉底哲学家，以提出宇宙的原子理论而闻名。他被认为是唯物主义者，相信一切都是自然法则所致。

13. **狄奥多西·格里戈罗夫斯基·杜布赞斯基**（1900—1975），乌克兰裔美籍遗传学家和进化生物学家。他后来到美国工作，是进化生物学领域和统一现代进化综论的核心人物。

14. **古斯塔夫·海因里希·西奥多·艾米尔**（1843—1898），德国动物学家。被誉为生物学术语"直生论"的推广者，这一术语顾名思义，指的是向特定方向发生的进化。

15. **查尔斯·萨瑟兰·埃尔顿**（1900—1991），英国动物学家和生态学家，因其对现代人口生态学的研究而闻名。

16. **罗纳德·费舍尔**（1890—1962），英国统计学家，将孟德尔遗传学纳入到了达尔文理论，形成了现在称之为现代进化综论的研究。费舍尔还是后来被称为人口遗传学的重要创始人之一，尽管不太受欢迎，但他还是一位杰出的优生学家。

17. **弗朗西斯·高尔顿**（1822—1911），英国人类学家和统计学家，曾经发明了指纹识别技术。他也是杰出的优生学家。他是查尔斯·达尔文的堂兄弟。

18. **埃蒂安·乔弗罗伊·圣-希莱尔**（1772—1844），法国博物学家，他确立了"统一组成"的原则，表明在所有的动物体内，都可以找到单一一致基本结构。

19. **布莱恩·凯里·古德温**（1931—2009），加拿大数学家和生物学家。他是理论生物学的奠基人之一，理论生物学是数学生物学的一个分支，使用数学和物理学的方法理解生物学中的变化过程。

20. **史蒂芬·杰·古尔德**（1941—2002），美国古生物学家、进化生物学家和科普作家；他对社会生物学决定论持反对态度。

21. **罗伯特·爱德蒙·格兰特**（1793—1874），苏格兰医生和解剖学家，曾在爱丁堡大学授课，教过达尔文，后来搬到伦敦，在那里设立了著名的格兰特动物学博物馆，并成为英国第一位动物学教授。他在对海洋无脊椎动物的研究中发现海绵实际上也是动物。

22. **威廉·（比尔）D. 汉密尔顿**（1936—2000），英国进化生物学家，他的理论包括在间接亲属（例如侄女）上所表现出的社会偏袒常常显示为利他主义，事实上，它增强了个体的适切性，使其更能偏爱亲属。他因创立了汉密尔顿法则闻名于世。在以理查德·道金斯为首的科学家基于基因的进化理论研究中，汉密尔顿也扮演着先驱者的角色。

23. **约翰·史蒂文斯·亨斯洛**（1796—1861），英国植物学家、牧师和地质学家。在剑桥大学，他推行了一种有利于培养独立思考和探究能力的教学方法，并成为青年查尔斯·达尔文进行研究的灵感源泉。

24. **金·希尔**，美国人类学家，他的研究重心是人类行为的进化生态学。他因认为人类具有独一无二性，并成功将人类与社会性联系起来而闻名。

25. **约瑟夫·道尔顿·胡克**（1817—1911），著名的植物学家兼探险家，曾任英国皇家植物园总监，是达尔文的挚友。

26. **詹姆斯·哈顿**（1726—1797），苏格兰地质学家，以他的均变理论而闻名。

27. **托马斯·亨利·赫胥黎**（1825—1895），解剖学家，因对达尔文《物种起源》中的观点进行的众所周知的辩护而被誉为"达尔文的斗牛犬"。他撰写了《关于人在自然界中位置的证据》（1863）一书，书中明确地指出人类和猿类都是源自一个共同的祖先，人类和其他动物一样，也在进化，因此自然选择对人也同样在起作用。

28. **弗莱明·詹金**（1833—1885），英国工程师，以发明缆车而闻名。

29. **史蒂夫·琼斯**（1944 年生），威尔士遗传学家和进化生物学家，在伦敦公共研究大学任职。

30. **斯图尔特·艾伦·考夫曼**（1939 年生），美国理论科学家、训练有素的医学博士。他最著名的论点是：生物系统和生物体的复杂性可能源于达尔文提出的自然选择学说，也可能源自自我组织系统。

31. **木村茂子**（1924—1994），日本生物学家，创立了分子进化的中性理论，表明大多数物种变化的原因在于漂变而不是自然选择。

32. **查尔斯·金斯利牧师**（1819—1875），英国神职人员，不仅口头盛赞并撰写文章支持达尔文的思想和进化理论。

33. **阿斯特里德·科德里克·布朗**，美国生态学家和进化生物学家，专注于淡水鱼的行为研究，特别是配偶识别系统的进化及其在物种形成中的作用。

34. **让-巴蒂斯特·拉马克**（1744—1829），法国生物学家，他在 1809 年所著的《动物学哲学》中指出了物种是可变的这一概念，该书还指出物种的变化可以世代继承。

35. **理查德·莱万廷**（1929 年生），进化生物学家和遗传学家，以与史蒂文·罗斯及莱昂卡·米恩合著的《不在我们的基因里：生物学、意识形态和人性》一书而闻名。

36. **卡尔·林奈**（1707—1778），瑞典植物学家，他发明了一种将所有生物体（包括人类）分类为子集的系统。在他 1735 年的《自然系统》一书中，他还发明了被称为二项命名法的分类命名系统（即每种生物用两个名字来表示，如智人，Homo sapien），沿用至今。林奈主张物种（和生物体）是固定不变的。

37. 约翰·卢伯克（1834—1913），数学家、科学家、银行家、政治家，也是威尔特郡阿维伯里第一男爵。和达尔文一样，他也住在肯特郡的达温村，是达尔文儿时最要好的伙伴之一。

38. 查尔斯·莱尔（1797—1875），地质学家、均变论倡导者。该理论认为今天改变地球的力量在过去也是一样。莱尔不仅从专业上，也在个人关系上给予达尔文极大的支持。

39. 托马斯·罗伯特·马尔萨斯（1766—1834），英国人口统计学家、神职人员。他提出一项理论：人口受到疾病和饥荒的制约。达尔文和华莱士在阅读马尔萨斯于 1798 年首次发表的《人口学原理》之后，分别提出了各自的自然选择学说。

40. 哈里特·马蒂诺（1802—1876），社会理论家、著名的辉格党员，撰写了 35 本著名的社会理论著作。她还与达尔文的哥哥伊拉斯谟斯有过恋爱关系。

41. 恩斯特·迈尔（1904—2005），德国生物学家、诺贝尔奖获得者，大力发展了生物物种概念，被认为是现代进化综论的奠基人之一。

42. 格雷戈尔·孟德尔（1822—1884），曾是摩拉维亚僧侣，他在豌豆杂交实验的基础上，于 19 世纪 60 年代撰写了关于遗传学和遗传特征的文章，这些文章直到 1900 年才得到广泛地阅读，填补了现代进化理论中的空白。

43. 杰弗里·米勒（1965 年生），进化心理学家，现于新墨西哥大学工作。他的研究兴趣主要是人类择偶研究。

44. 乔治·爱德华·摩尔（1873—1958），英国一位有影响力的现实主义哲学家，他对伦理问题采取系统化的方法进行研究，一丝不苟，兢兢业业，在哲学中成功创立了"分析"传统。他还因为伦理非自然主义和自然主义哲学辩护而著名。

45. 刘易斯·亨利·摩根（1818—1881），美国人类学家，达尔文同时代人，他的各种社会的技术"进化"理论流传了好几代，直到受到了越来越多的相对主义人类学家发起的挑战。

46. 艾萨克·牛顿（1642—1726），英国物理学家，他发现了重力原理及运动定律。

47. **弗里德里希·尼采**（1844—1900），德国哲学家，因提出了超人等哲学概念以及他在宗教有用性和意义方面的富有争议性的著作而闻名。

48. **诺亚**，创世纪圣经中提到的诺亚方舟故事中的主要人物，故事讲述了上帝如何发动洪水摧毁地球上所有的生物；每类动物只有两只被诺亚船或方舟救起。

49. **理查德·欧文**（1804—1892），英国古生物学家和比较解剖学家，可能因为创造"恐龙"一词而最闻名。他直言不讳地反对达尔文的自然选择进化论观点——尽管他同意进化客观存在，但他相信其驱动机制要比达尔文所提出的复杂得多。

50. **威廉·佩利主教**（1743—1805），英国神职人员和哲学家。他在《自然神学：从自然现象中搜集的关于神性存在和其属性的证据》一书中利用现在著名的钟表匠类比指出，上帝的存在被自然世界的美丽和复杂所证实。

51. **史蒂文·平克**（1954 年生），加拿大美国心理学家，以《空白的石板：现代对人性的否定》一书而闻名，此书对人类行为采取了生物决定论的立场（也就是说，他认为我们的行为取决于我们的生物构造特点）。

52. **柏拉图**（公元前 428—348），古希腊哲学家，他主要关注与正义、美和平等相关的话题。他在雅典成立了一所名叫"学院"的学校，是西方世界第一所高等教育机构。

53. **让-雅克·卢梭**（1712—1778），瑞士作家和哲学家。他撰写了《社会契约》（1762 年），挑战了国家或宗教凌驾于个人之上的做法。

54. **亚当·塞奇威克**（1785—1873），英国著名的地质学家，也是达尔文在牛津大学时的地质课老师，但他对学生达尔文的进化论持反对态度。

55. **约翰·梅纳德·史密斯**（1920—2004），英国进化生物学家，因其对在进化背景下进行的博弈论研究而闻名。

56. **赫伯特·斯宾塞**（1820—1903），英国生物学家和人类学家，达尔文同时代人，因创造"适者生存"这一术语而闻名。

57. **沃尔特·萨顿**（1877—1916），美国遗传学家，以格雷戈尔·孟德尔的研究成果为出发点，他提出了一个关于染色体的重要理论。

58. **弗雷德里克·坦普尔**（1821—1902），英国牧师，后任坎特伯雷大主教。因对宗教与科学的相互作用关系颇感兴趣而闻名。

59. **迈克尔·托马塞洛**（1950 年生），出生于美国的心理学家，现居住在德国。他重点研究人类智力的起源。

60. **雨果·德弗里斯**（1848—1935），荷兰植物学家和遗传学家。在 19 世纪 90 年代，在重新发现格雷戈尔·孟德尔关于豌豆植物中遗传定律后，提出了基因的概念。他还因提出突变这一术语，并且创立了基于基因突变的进化论而闻名。

61. **阿尔弗雷德·拉塞尔·华莱士**（1823—1913），威尔士博物学家，独立提出了一种自然选择学说，与达尔文的物种演变机制理论大体相同。达尔文和华莱士被公认为共同发现了自然选择这一概念。

62. **乔西亚·韦奇伍德**（1730—1795），英国陶器设计师和制造商。他用科学的方法制作陶器，并因对材料方面的深入研究而闻名。

63. **弗里德里希·利奥波德·奥古斯特·魏斯曼**（1834—1914），德国进化生物学家。他因提出种质理论而闻名。该理论指出，在多细胞生物体中，遗传只能通过生殖细胞——配子或卵子和精子细胞得以实现。这种思想对现代进化综论至关重要。

64. **爱德华·奥斯本·威尔逊**（1929 年生），美国生物学家和环境活动家，被认为是社会生物学（对社会行为的生物学本质进行的研究）的主要思想家和理论家。

65. **玛琳·祖克**（1956 年生），美国生物学家，其研究重点是性别选择和配偶选择，以及次生性征的演变。

WAYS IN TO THE TEXT

- Charles Robert Darwin (1809–82) was a British naturalist* (a scholar of the natural world) and geologist* (a scholar of the formation and history of the earth's rocks).

- *On the Origin of Species by Means of Natural Selection* explains the origin, succession, and extinction* of species by "natural selection"* (the process by which the organisms best adapted to a particular environment out-compete the organisms that are less well adapted, and successfully pass on their useful adaptations to the next generation).

- The main concepts in *Origin* are central to the life sciences.

Who Was Charles Darwin?

Charles Darwin, the author of *On the Origin of Species by Means of Natural Selection* (1859), a work explaining evolution* by the process he called "natural selection," was a renowned natural historian*. His extraordinarily influential book changed the way scientists understand the origin and development of life on earth, and earned its author the title "father" of evolutionary theory.

Darwin was born in 1809 in the market town of Shrewsbury in England. As part of the well-known Darwin-Wedgwood family, he was born into both wealth and privilege. In 1831, he graduated from Christ's College, Cambridge, with a BA; the following winter he became ship's naturalist aboard the HMS *Beagle**, a Royal Navy vessel on a five-year scientific trip around the globe. On Darwin's return, his father organized investments that allowed him to be a "self-funded gentleman scientist."[1]

Darwin's own social circle was one of scientific academics and radical thinkers such as the geologist Charles Lyell* and the social theorist Harriet Martineau*, who helped him to develop his theory of natural selection. Darwin initially delayed publication of his theory, fearing the public outrage that would follow; this was because it seemed to refute Christian teachings regarding the origins of life on earth (among other things). In 1858, however, the naturalist Alfred Russel Wallace* wrote to Darwin proposing a similar theory for the mechanism of evolution. This forced Darwin's hand, and he decided to publish immediately.

Wallace (1823–1913) was a Welsh naturalist, collector, and tropical field biologist (a specialist of tropical organisms). In 1858, he published the theory of evolution by natural selection together with Charles Darwin in an article presented to the Linnean Society* (an institution founded in London for the discussion and study of natural history). The following year, Darwin published a more detailed account of his theory in On the Origin of Species. In total, he published 25 books on diverse topics such as barnacles, plants, and earthworms. Having devoted his life to scientific research, he died in the county of Kent at his family home, Down House, in April 1882. He is buried at Westminster Abbey, in London.[2]

What Does On the Origin of Species Say?

The main subject of On the Origin of Species is evolution— the process by which all earth's species have descended from a common ancestor. The theory of natural selection, explaining how organisms adapt to their environment to ensure their survival, is based on the observation that while plants and creatures tend to over-reproduce, resources are finite. Yet population sizes generally remain stable.

To account for this, Darwin points out that there are fine variations between organisms, notably in their hardwired behavior (what he calls "instinct"). As a species* is made up of more individuals than can possibly survive all together, there is a struggle between them for existence. In this competition for limited resources, those individuals best adapted to their environment are more likely to survive and pass on their traits (the differences in their behavior and physical attributes). Gradually, a species evolves to occupy a different niche, so becomes a new species, and is unable to reproduce with the species from which it descended.

The concept of natural selection remains a major part of the study of evolution, along with mutation* (a relatively sudden change in an animal's genetic makeup, leading to a change in its behavior or physical constitution), and genetic drift* (roughly, the process by which genetic information changes over time as animals reproduce).

Secondary themes in *On the Origin of Species* include sexual selection,* speciation* (the process by which a new species evolves), and gradualism* (the way in which species change in intermediate stages over time).

Sexual selection (natural selection via the choice of a mate, for example) can take place either within the sexes or between them. In intrasexual selection*, members of the same sex compete for mates of the opposite sex. This is mostly seen in competition between males: the male with the best fighting technique, largest body size, or the biggest weapons will have the highest chance of winning. The winner then gains exclusive access to mates, and so out-reproduces the losers; natural selection occurs if the characteristics that determine the outcome of the contest are inherited. In intersexual selection, *sometimes referred to as "female choice," members of one sex choose among potential mates based on certain qualities. If the preferred individuals are genetically different from their rivals, natural selection occurs.

Darwin saw evolution as a slow and gradual process. In *Origin*, he introduces the concept of gradualism—species evolving and accumulating small variations over long periods of time. He explains, too, how a new species can arise by way of population speciation, which occurs when a population divides and two new species develop. If a species occupies a large area, variations in the environment will mean that various individuals experience different pressures, and so adapt differently.

Within separate regions, new species arise because natural selection acts independently in each environment. Darwin suggests that there must also be hybrid zones between regions; the individuals within them, being less well adapted to the conditions of either adjoining zone, eventually become extinct.

While evolution generally takes place over a very long time, there are exceptions; for example, some viruses and insects can become resistant to certain pesticides over relatively short periods.

Why Does *On the Origin of Species* Matter?

On the Origin of Species is one of the most influential scientific texts ever written. The idea of natural selection proposed within it provided an explanation both for the evolution of complex organisms and the exquisite fit of those organisms to their environment, without the need for divine intervention. Darwin used extensive evidence to support his theory of descent with modification*—the pattern of evolution—and the idea that all living organisms are related by way of a common ancestor.

Perhaps even more remarkable was that Darwin wrote the book in such a way that both scientists and the general public could understand his theory. This wide readership resulted in controversy, however, as the text was regarded as a challenge to the prevailing ideology of state and Church.

In establishing the theory of evolution, Darwin laid the foundations for the field of evolutionary biology*. To a lesser extent, he had an influence on disciplines as diverse as theology* (the systematic study of religious ideas, commonly conducted through scripture), modern philosophy, oceanography* (the scientific study of oceans), linguistics* (the study of the structures and nature of language), anthropology* (the study of human cultural and social life), history, and economics.

Nearly 160 years after the publication of *Origin,* evolution is still a contentious issue. Many remain skeptical, and fewer than half (48 percent) the people in the United States accept Darwin's theory of natural selection.[3] At the time, atheists* (those who do not believe in any god) and materialists* (those who look for material causes alone for all physical phenomena) thought it represented a serious, decisive challenge to religion. Evolution and natural selection justified dismissing the idea that a deity may have played a role in creating humans and animals.

Religious believers, however, who saw the wonders of the natural world as God's creation, regarded the theory as a great insult. Referencing positivism*, the principle that knowledge can only advance through scientifically verifiable facts, the evolutionary biologist Ernst Mayr* wrote: "Eliminating God from science made room for strictly scientific explanations of all natural phenomena; it gave rise to positivism, and produced a powerful intellectual and spiritual revolution, the effects of which have lasted to this day."[4] Even today, religious sentiment is mixed. Some accept the concept of natural selection but others still oppose Darwin's theory.

While there have been countless developments in evolutionary theory since the publication of *Origin*, the theory of natural selection still holds as a way of understanding species change;[5] and Darwin's ideas have come to influence both science and society.

---

1. Janet Browne, *Charles Darwin: A Biography. Vol. 1: Voyaging* (Princeton: Princeton University Press, 1996), 434–5.

2. Janet Browne, *Charles Darwin: A Biography. Vol. 2: The Power of Place* (London: Pimlico, 2003), 496.

3. Pew Forum, "Religious Groups: Opinions of Evolution," February 4, 2009, accessed February 5, 2016, http: //www.pewforum.org/2009/02/04/religious-differences-on-the-question-of-evolution/.

4. Ernst Mayr, "Darwin's Influence on Modern Thought," *Proceedings of the American Philosophical Society* 139, no. 4 (Dec 1995): 317–25.

5. H. Allen Orr, "Testing Natural Selection with Genetics," *Scientific American* 300, no. 1 (2009): 44.

SECTION 1
INFLUENCES

MODULE 1
THE AUTHOR AND THE
HISTORICAL CONTEXT

KEY POINTS

* Evolutionary* theory and the mechanism of natural selection*
  together provide the framework for all contemporary biological
  sciences.

* Charles Darwin's social standing and his proximity to the leading
  intellectual minds of his time facilitated the development of his
  theory.

* Mid nineteenth-century Europe was bursting with post-
  Enlightenment* reformist ideas concerning slavery, colonialism,
  and the nature of man. (The Enlightenment emphasized notions
  of rationality and liberty; colonialism is the process by which one
  nation or people exploit another through the occupation of land.)

Why Read This Text?

Charles Darwin's *On the Origin of Species by Means of Natural
Selection* (1859) explains how species have evolved from a common
ancestor through the process of natural selection. Darwin drew on
different sources to formulate a theory that would revolutionize
science, at a time when the advancement of knowledge was restricted
by religious dogma. Afterward, evolution became a legitimate field of
scientific inquiry.[1]

Darwin's prose is clear and personal. Despite the controversy
it caused, *Origin* never argues against the existence of a deity,
only against the notion of intelligent design* (the belief that life
on earth was created by an intelligent god) as an explanation for

the origin of species.[2] Many contemporary scientists would argue that science and religion are not in opposition, as they address two very different questions.[3] Religion seeks to explain the meaning of individuals' existence and provides moral guidance; science seeks to understand life on earth as a whole, and places no value judgment on characteristics or behavior.

The text is relevant to studies of humans, animals, and plants,[4] offering a theory that still serves as a valid explanation for species change.[5] It is central to studies such as biology, evolutionary psychology* (the study of the human mind and behavior conducted in the light of evolutionary theory), and botany* (the study of plants). To a lesser extent, it also influenced theology* (the systematic study of religious ideas, commonly conducted through religious scripture), modern philosophy, oceanography* (the scientific study of oceans), linguistics* (the study of the structures and nature of language), anthropology* (the study of human cultural and social life), history, and economics.

> *"Is man an ape or an Angel? I, my Lord, am on the side of the angels. I repudiate with indignation and abhorrence those new fangled theories."*
> —— Benjamin Disraeli, speech, 1864, quoted in *Charles Darwin: The Man and His Influence* by Peter Bowler

Author's Life

Charles Robert Darwin was born in 1809 to an affluent, intellectual family. In an effort to give him a respectable occupation, his father sent him to the University of Edinburgh Medical School at the age of 16. Darwin had no desire to be a doctor. He found the curriculum dull and the sight of operations performed without anesthetic disturbing. Instead, he became interested in natural history (the natural world), reading widely, collecting specimens, and dissecting small animals. He befriended a taxidermist, who taught him how to skin and stuff birds, and the zoologist* Robert Grant*, who introduced him to the French biologist Jean-Baptiste Lamarck's* theory of transmutation* (a first attempt to explain the idea that changes can be passed from generation to generation).[6] He joined the Plinian Society*—a club for students interested in natural history—and took part in debates on science.

Darwin's neglect of his medical studies annoyed his father and so, in 1827, he was sent to Christ's College, Cambridge, to study theology. But Darwin's interest in natural history continued. He took lectures in botany and went on long plant-collecting expeditions with the clergyman and botanist John Stevens Henslow.* Darwin also developed an interest in geology* (inquiry into the formation and history of the earth's rocks) and attended a course on the subject taught by the geologist Adam Sedgwick*.

In 1831, after completing his studies at Cambridge, Darwin took up a post as ship's naturalist and gentleman companion to Captain Robert Fitzroy on board the HMS *Beagle*.* Darwin had to pay his own way and the trip lasted longer than he had expected—five years instead of two. During the voyage, he collected numerous specimens, dissecting some and stuffing others. He also spent a great deal of time surveying the scenery and found evidence to support the theory of geological uniformitarianism*—the idea that the earth has experienced a series of physical changes over time by means of natural processes.

He returned in 1836, becoming an instant celebrity after publishing his journal as a travel book, *Journal of the Voyage of the Beagle*. He earned the respect of his father and his peers, and was elected secretary of the Geological Society. During the same period, he struck up friendships with the biologist Thomas Henry Huxley* and the geologist Charles Lyell,* people who would be invaluable to his thought and career.[7]

In 1839, Darwin married his cousin Emma Wedgwood.* Less than a year later, he became ill and the couple moved to Down House in Kent. Perhaps this illness gave him the excuse he was looking for to withdraw from the distractions of society and concentrate on his scientific research.[8] Darwin had ten children, three of whom died in childhood; he was as devoted to his family as to the natural sciences. After the publication of *Origin*, poor health kept him away from the ensuing heated public debates, but he was kept up to date with events by letter correspondence with his close friends, such as Huxley. In Kent, he lived a quiet life, rarely socializing and instead focusing on family life and writing books and scientific papers.

Author's Background

In his youth, Darwin collected minerals, birds' eggs, and insects—pursuits regarded by his father, the scientist and medical doctor Robert Waring Darwin*, as a self-indulgent waste of time. There were several other scientists in the family, among them his grandfather Erasmus Darwin*, a physician and naturalist, and his cousin Francis Galton*, a statistician and anthropologist. His maternal great-grandfather was the famous entrepreneur Josiah Wedgwood,* a potter and prominent abolitionist*—an activist in the struggle to abolish slavery. Both families were largely Unitarians—belonging to a sect of Protestantism, one of the two largest branches of the Christian faith. Darwin's dislike of slavery and his support for the Great Reform Act of 1832* (legislation that granted many previously disenfranchised men the right to vote) reveal his general tendency toward liberal* and reformist views.

With personal financial security, and therefore more leisure time for exploration than the average Victorian, Darwin was able to pursue his interest in natural history. His social standing gave him access to some of the greatest scientific, philosophical, and literary minds of the day,[9] including the scientist and mathematician John Lubbock* and the social theorist Harriet Martineau*. In this environment, his ideas blossomed, backed up by his practical experiments in botany, zoology (the study of animals), and biological inheritance* (the various characteristics that can be passed from parent to child). At the time of the text's publication in 1859, Darwin was given further professional and personal support by Charles Lyell, Thomas Huxley, and the botanist Joseph Dalton Hooker*. Despite the religious controversies surrounding his theory of evolution,[10] Darwin maintained an academic career of note to the end of his life.

1. Edward J. Larson, *Evolution: The Remarkable History of a Scientific Theory* (New York: Modern Library, 2004).

2. Mark Ridley, *How to Read Darwin* (London: Granta Books, 2006).

3. Steve Jones, *Darwin's Island: The Galapagos in the Garden of England* (London: Little Brown, 2009).

4. Cynthia Delgado, "Finding the evolution in medicine," *National Institutes of Health Record*, 58.15 (2006), 1–8.

5. H. Allen Orr, "Testing Natural Selection with Genetics," *Scientific American* 300, no.1 (2009): 44.

6. Janet Browne, *Charles Darwin: A Biography. Vol. 1: Voyaging* (Princeton: Princeton University Press, 1996), 75–6.

7. Browne, *Voyaging*, 355–6.

8. Janet Browne, *Charles Darwin: A Biography. Vol. 2: The Power of Place* (London: Pimlico, 2003), 5.

9. Browne, *The Power of Place*, 5.

10. Browne, *The Power of Place*, 5.

MODULE 2
ACADEMIC CONTEXT

KEY POINTS

* The natural sciences are concerned with understanding the physical world.

• During the intellectual movement known as the Enlightenment* of the seventeenth and eighteenth centuries, in which society and scholarship took a turn toward rationality, several scientists had considered the development of new species but were unable to explain fully how it occurs.

* *On the Origin of Species* answers the then-pressing question of how species change gradually over time by proposing a mechanism called "natural selection."*

The Work in its Context

It would be difficult to think of a book that embodies the Victorian era more fittingly than Charles Darwin's *On the Origin of Species by Means of Natural Selection* (1859). In the nineteenth century, England was bursting with ideas and social innovations that challenged both state and Church; among them were the Great Reform Act of 1832,* which granted the vote to many thousands of men previously denied it, debates concerning man's place in nature, and the abolitionist* (anti-slavery) movement. The blooming of new social and scientific ideas, nonconformist* sects (religious faiths that confronted the Anglican Church), and the anti-alcohol temperance* movements combined to challenge the post-Enlightenment world and its dominant ideologies.

There had already been rumblings on this front, with European and American calls to give the vote to all men (universal male suffrage) and the mid-seventeenth century emergence of liberalism* (the protection of an individual's liberty by a system of checks and balances on government). Ideas as subversive as those found in *On the Origin of Species*, which suggests that nature, not God, created life, challenged authority at the highest imaginable level. The work was perfectly placed to question the dominant paradigm* (the model by which knowledge was attained and understood): the idea that organisms are static and never change, and were created by a divinity who also gave human beings dominion over them. Many scientists had long wanted to challenge this model—but few had dared.

> *"When we contemplate every complex structure and instinct as the summing up of many contrivances, each useful to the possessor, nearly in the same way as when we look at any great mechanical invention ... when we thus view each organic being, how far more interesting, I speak from experience, will the study of natural history become!"*
>
> —— Charles Darwin, *On the Origin of Species by Means of Natural Selection*

Overview of the Field

According to Christian belief, the earth was shaped by the damage inflicted on it when God punished man with a great flood; the earth is a static ruin and has not changed ever since. The species that exist today have always existed, and their extinction or change is inconceivable.

Christian doctrine was not the only important intellectual influence. The philosophy of the ancient Greek thinker Plato* had a similar effect. According to Plato, the universe contains fixed ideal types of everything that can be seen and felt (or indeed imagined); these ideal types are hidden from us here in the everyday world, where only their distorted variations are visible. When this thinking, termed essentialism*, is applied to nature, biological change becomes impossible: although wild elephants, for example, might vary in height, contained hidden within each one lies a blueprint of the ideal elephant[1] and "change" is nothing but a distorted variation.

This view was called into question in the eighteenth century, when geologists began to discover certain types of shellfish that had existed in the past but were no longer alive—a challenge to the biblical view of the world. Such scientists continued to find more extinct life forms. In order to reconcile these findings with religious orthodoxy, Christians felt they must be the result of intermittent catastrophes: there had not been just one flood but many, the last being Noah's*.

By the end of the eighteenth century, evidence in favor of evolution was emerging in other disciplines too. The year before *On the Origin of Species* was published, an almost complete skeleton of the dinosaur *Hadrosaurus* was found in the US state of New Jersey. A decade earlier, in 1841, the British naturalist Richard Owen* had presented dinosaurs as a separate taxonomic* group (an order of animals unique to itself), coining the name *Dinosauria*.[2] Darwin, in fact, left dinosaurs out of his text, since only a few specimens had then been found; no one knew anything about them or if they had left any living descendants. Stronger supporting evidence for evolution came from comparative anatomy*—the study of the comparison and contrast of the anatomies of different species.

Many vertebrate animals—animals with a spine—have a very similar anatomical structure. So the same five-digit structure is observed in the human hand as in the wing of a bat, which has five digits ("fingers") still present in its anatomy. Furthermore, comparative anatomists observed that there are phases in the development of the human embryo that look identical to those of birds, reptiles, and mammals. All this evidence was inconsistent with the notion of intelligent design* (that there is a divine "designer" of the earth's organisms), and Darwin would come to adopt it as supporting evidence for common descent.

Academic Influences

There were many scholars who influenced Darwin, and whose ideas appear in *On the Origin of Species*; in particular, the geologist Charles Lyell*, who developed the theory of uniformitarianism*—the idea that the earth has undergone a series of physical changes by means of natural processes over geological time. Darwin thought that if the physical environment could change, then so could animals and plants. Indeed, they had to adapt to fit the changed environment, or become extinct.

The French biologist Jean-Baptiste Lamarck* was also of key importance. Devoted to the doctrine of the Great Chain of Being*—an arrangement of natural types ordered from the simplest to the most complicated—Lamarck saw the idea as a staircase that species ascend, striving to become more complex. He theorized that individuals could modify their bodies in response to environmental problems and pass these new beneficial traits on to their offspring.[3]

Another major influence was the pioneering statistician Thomas Robert Malthus*, author of *An Essay on the Principle of Population*, first published in 1798. This text inspired Darwin to consider the possibility of a natural mechanism by which overpopulation is held in check.

Darwin's theory of natural selection could not have existed without the work of naturalists*, anatomists* (specialists in animal anatomy), and taxonomists* (specialists in the categorization of organisms), who helped him to catalogue and describe the vast hoard of specimens he had collected during his voyage on the HMS *Beagle**. The anatomist Richard Owen* and the ornithologist John Gould* (an expert in birds) described some of these specimens. It was Gould who realized that birds brought back from the Galapagos Islands that Darwin believed to be blackbirds, grosbeaks, and finches, were in fact 12 separate species of finches.[4]

---

1. Edward J. Larson, *Evolution: The Remarkable History of a Scientific Theory* (New York: Modern Library, 2004).

2. Deborah Cadbury, *Terrible Lizard: The First Dinosaur Hunters and the Birth of a New Science* (New York: Henry Holt, 2000).

3. Rebecca Stott, *Darwin's Ghosts: In Search of the First Evolutionists* (London: Bloomsbury, 2012), 194–7.

4. Adrian Desmond and James Moore, *Darwin* (London: Michael Joseph, 1991), 209.

MODULE 3
THE PROBLEM

KEY POINTS

* Organisms change and adapt in nature according to their specific environment and the organisms they are descended from.

* Biblical literalists (those who take the explanations offered by the Bible literally) disagreed; some esteemed scientists agreed with them.

* Darwin proposed that species do indeed change over time, and set out to discover the mechanism that underpins such changes.

Core Question

On the Origin of Species by Means of Natural Selection by Charles Darwin explains how all life on earth has evolved by the process of natural selection*.

At the core of Darwin's investigation are the questions as to why some animals and plants have continued to exist for long periods of time while others have become extinct, and why there are similarities between many of these extinct forms and present-day species.

In the nineteenth century, it was assumed that the world and its inhabitants had always been the same and that everything was created by God. After his voyage around the world on the *Beagle**, Darwin began to question this doctrine, having seen firsthand evidence in the geology of South America and its fossils that contradicted this belief. By 1837, he was convinced that life had evolved, and he wanted to know how these evolutionary* changes had taken place. Over time he would come to believe that new species come about not as the result of divine intervention but by adapting to changing conditions.[1]

> "How have all those exquisite adaptations of one part of the organization to another part, and to the conditions of life, and of one distinct organic being to another being, been perfected?"
> —— Charles Darwin, *On the Origin of Species by Means of Natural Selection*

The Participants

Christian doctrine claimed that the world had not changed since God created it, which placed a 6,000-year limit on the age of the planet. If the earth never changes physically, then there is no need for living things to alter either. In 1749, however, the French naturalist Georges-Louis Leclerc (later Comte de Buffon)* questioned this, arguing that life had a history of its own and that the earth was more than 6,000 years old.[2] Buffon also observed similarities between humans and other primates such as chimpanzees, and suggested that human beings and the other apes share a common ancestry. While he raised the idea of biological change, Buffon did not, however, provide a coherent mechanism for how such changes occur.[3]

The naturalist Jean-Baptiste Lamarck* presented another

challenge to the biblical view in his *Philosophie Zoologique* (1809), offering a theory of evolution based on the mutation* of species: his argument was that environmental challenges could force species to modify their bodies over time to gain personal advantage. He also suggested that offspring inherit these modifications from their parents[4]. In 1826, the British naturalist Robert Grant* began to speak publicly about evolution and promoted Lamarck's theory of transmutation* of species in Britain.[5]

In response to these challenges to creationism*, Bishop William Paley* argued that adaptations are the supernatural creation of God; the anatomist* Georges Cuvier also criticized Lamarck's ideas, insisting that species are immutable (unchanging, and unchangeable). He did not reject evolution for religious reasons, however, but for reasons to do with evidence—the fossil record did not show any striving toward perfection.[6] For Cuvier, catastrophic changes were followed by acts of spontaneous generation.

The Contemporary Debate

In the 1700s, geologists* began to discover rocks that contained within them records of previous extinction events (periods when large numbers of species die out simultaneously). Scientists at the time tried to accommodate Christian belief by introducing the theory of intermittent catastrophes, which posited that after each disaster, God had recreated the living forms on earth. The final catastrophe was Noah's* flood. Such explanations nevertheless failed to explain why some animal types had perished during these events, while others had survived: a badger species existed in the

Miocene Age, for example, that was almost identical to the badger of the present day.[7]

Building on the evidence from fossils found in ancient rocks, in 1788 the Scottish geologist James Hutton* introduced the theory of uniformitarianism*: the idea that the earth had undergone continuous physical change in the past and that the same transformative process is continuing to the present day. This change is constant and gradual rather than a series of catastrophic events. Hutton's theory was largely overlooked until the nineteenth century, when the geologist Charles Lyell* sought to develop and popularize it.[8]

The French biologist Jean-Baptiste Lamarck made a forceful argument in his *Philosophie Zoologique* (1809). Having absorbed the ideas of essentialism* from the thought of the Greek philosopher Plato, according to which any animal or physical object possesses a set of attributes necessary to its identity and function, and also that of the Great Chain of Being* (the idea that organisms are structured hierarchically, with those higher up superior to those below), Lamarck believed that all living things are engaged in a constant struggle to reach ever-greater complexity. Their ultimate goal is to become as complex as man. As some organisms become more complex, gaps appear at the bottom of the ladder, to be filled by simple, spontaneously generated organisms. Lamarck thought that two forces direct this process: the inherent drive toward complexity, and the environment. Living things, he believed, rise to ecological challenges by modifying their bodies, and these changes are passed on to the next generation. His argument was so persuasive that it remained the most influential evolutionary theory until that of Darwin and the Welsh naturalist Alfred Russel Wallace,* who came to very similar conclusions.[9]

However, Lamarck's theory of the transmutation of species was associated with the radical materialism* of the Enlightenment * ("materialism"here means the assumption that all physical phenomena have physical causes), and was greeted with hostility. In his youth Darwin had read Bishop's Paley's *Natural Theology* (1809), a book written partly in response to this theory. In it, Paley declares adaptations to be the supernatural creations of God.[10] Indeed, the existence of such adaptations in nature provides one of the main philosophical arguments for the existence of God. This is known as the argument for "providential design".

Lyell also criticized Lamarck's theories in *Principles of Geology* (1830, 1833). Although Lyell wrote about uniform change in inorganic matter, he refused to believe in the possibility of biological change. Instead, he proposed that each species has its "centre of creation" and is designed for a particular environment; species become extinct when the environment that supports them changes.[11]

1. Edward J. Larson, *Evolution: The Remarkable History of a Scientific Theory* (New York: Modern Library, 2004).

2. Rebecca Stott, *Darwin's Ghost: In Search of the First Evolutionists* (London: Bloomsbury, 2012).

3. Georges-Louis Leclerc, Comte de Buffon, *Les Époques de la Nature*, in *Histoire Naturelle, générale et particulière, avec la description du Cabinet du Roi* (Paris: imprimerie nationale, 1749–1788).

4. J. B. Lamarck, *Zoological Philosophy: An Exposition with Regard to the Natural History of Animals*, trans. Hugh Elliot (Chicago: University of Chicago Press, 1984).

5. Stott, *Darwin's Ghost*.

6. G. Cuvier, *Tableau elementaire de l'histoire naturelle des animaux* (Paris: Baudouin, 1798), accessed February 16, 2016, https: //archive.org/details/ tableaulment00cuvi, 71.

7. Robert Chambers, *Vestiges of the Natural History of Creation* (London: John Churchill, 1844).

8. Larson, *Evolution*.

9. Larson, *Evolution*.

10. William Paley, *Natural Theology: or, Evidences of the Existence and Attributes of the Deity* (London: J. Faulder, 1809).

11. C. Lyell, *The Principles of Geology*, Vol 2 (London: Murray, 1830–3), chapter 2.

MODULE 4
THE AUTHOR'S CONTRIBUTION

KEY POINTS

* Scientists had long been looking for a working model to explain variation within and between species. It was Darwin (and his contemporary Alfred Russel Wallace)* who hit on it at last.

* Natural selection* provided a working mechanism to explain how different species come about and how they change over time.

* Darwin proposed that individual organisms vary in their characteristics, and that they compete with each other in a struggle for existence in a world of limited resources.

Author's Aims

In *On the Origin of Species by Means of Natural Selection*, Charles Darwin's aim was to explain adaptation and evolutionary* change. Such a theory needed to account for the full diversity of life on earth and explain why organisms look and behave the way they do. The adaptations included physical changes in a part of the body, such as a wing, as well as a particular behavior—both being designed to improve an organism's chances of survival and reproduction. Such adaptations could not arise by chance, and they required an explanation.[1]

The concept of evolution has been around since the age of ancient Greece. When Darwin came to publish *On the Origin of Species*, many philosophers and scientists had already accepted that animal and plant life evolves over time.[2] By the nineteenth century, the disciplines of geology,* embryology, and anatomy* had accumulated plenty of evidence in favor of evolution ("embryology" here refers to the study of the development of animal embryos in the womb). Scientists had attempted to explain how the various forms and varieties of animal and plant types had come about, but no one had managed to do so convincingly. Darwin changed this, offering a coherent argument understandable to scientists and nonscientists alike.

> "Light will be thrown on the origin of man and his history."
> ——Charles Darwin, *On the Origin of Species by Means of Natural Selection*

Approach

In *On the Origin of Species*, Darwin constructs a scientific argument using the comparative method—he presents the reader with a set of detailed observations and then uses reasoning to support his idea. In order to illustrate how varieties occur in nature, for instance, he uses examples from domestic animals. He explains how the selective breeding of dogs has resulted in various breeds that all look very different from one another, even though they are all descendants of a common ancestor. This shows that a handful of breeds can produce hundreds more—which, Darwin concludes, must also be possible in wild animals.

Darwin provides evidence, too, for his idea of gradualism*. Using eyes as an example, he explains how the eyes of mammals* might have started out as simple light-sensitive organs, as seen in invertebrates (animals without a spine).³ Of course, as eyes are composed of soft tissue they do not appear in fossils, but Darwin argues his case in principle, inferring the stages of evolution that led to the human eye. In this way, he rejects the notion of intelligent design* and replaces supernatural intervention with natural selection.⁴

Contribution in Context

Darwin and his contemporary, the naturalist* Alfred Russel Wallace, conceived an original theory for evolution that used evidence from the fields of geology, embryology, and anatomy. This was a quite a departure from the approach taken by Jean-Baptiste Lamarck* and some other German scientists, who built on speculation and theorizing. It is also worth noting that both Darwin and Wallace started from human beings and worked back to animals when creating their theory of natural selection.

Although Darwin and Wallace are credited as co-discoverers of natural selection, a careful reading of the article they jointly published in 1858 shows they had different approaches to the theory. Only Darwin understood that competition is greatest within members of the same population, rather than between species. He also noted that species with a recent common ancestor tend to appear more similar, while those with a more distant common ancestor have fewer shared characteristics.⁵ This led him to his principle of divergence*, and how species evolve over time.

Wallace, on the other hand, emphasized the importance of the environment in shaping species.⁶ He believed that food supply and predation are particularly influential on population growth, and concluded, "The numbers that die annually must be immense; and as the individual existence of each animal depends upon itself, those that die must be the weakest—the very young, the aged, and the diseased—while those that prolong their existence can only be the most perfect in health and vigor—those who are best able to obtain food regularly, and avoid their numerous enemies."

Contrary to Darwin, Wallace emphasized that an individual is deemed *weak* not on the basis of biological inheritance, but pure chance; the example he gives is the very old or young.

---

1. Mark Ridley, *How to Read Darwin* (London: Granta Books, 2006).
2. Edward J. Larson, *Evolution: The Remarkable History of a Scientific Theory* (New York: Modern Library, 2004).
3. Ridley, *How to Read Darwin*.
4. Ridley, *How to Read Darwin*.
5. Michael Bulmer, "The theory of natural selection of Alfred Russel Wallace FRS," *Royal Society Journal of the History of Science* 59, no. 2 (2005): 125–36.
6. Bulmer, "The theory of natural selection," 125–36.

SECTION 2
IDEAS

MODULE 5
MAIN IDEAS

KEY POINTS

* The main theme explored by Darwin is that individuals vary within any population.
* This variation results in differing rates of survival and reproduction.
* Individuals with traits most favorable to survival live, go on to reproduce, and pass on those traits to their offspring.

Key Themes

The core themes of Charles Darwin's *On the Origin of Species by Means of Natural Selection* are variation, biological inheritance*, and competition within populations for resources. Together, they form the mechanism of evolution*, which Darwin termed "natural selection."*

In biology, there are two kinds of variation. The first includes any difference between cells, individual organisms, or groups of organisms as a result of genetic* differences (differences in genes,* which decide an organism's properties). This is known as "genotypic variation*." The second takes into account the influence of the environment on genetic potential; this is known as phenotypic variation*. Variations can be observed either in physical appearance, such as height, or in behavior. "Biological inheritance" describes the process by which genes are passed on from parent (or ancestor) to offspring.

Finally, "competition" in biology is an interaction between

two or more organisms that require the same limited resources (such as food, water, or a mate, which the organisms require in order to grow, survive, and reproduce). As an organism cannot secure a resource if another organism has already consumed or defended it, competitors reduce the potential of others for growth, reproduction, or survival.

> "We are not here concerned with hopes or fears, only with the truth as far as our reason allows us to discover it."
>
> —— Charles Darwin, *The Descent of Man, and Selection in Relation to Sex*

Exploring the Ideas

Darwin begins *On the Origin of Species* by illustrating that there is built-in variation within species, using everyday examples from among domestic animals and plants. He observes that some domestic animals have an "extraordinary tendency to vary"[1]— notably the dog. Over several pages, Darwin describes the various dog breeds and their physical differences, speculating that they probably originated from just a handful of wild species. Darwin calls them varieties because it is possible to interbreed domestic dogs (*Canidae*), so they are classed as one species. From domestic animals he goes on to explore the natural variation that occurs in wild animals and plants. He admits that the domestic dog is a product of human manipulation—in the form of selective breeding, according to which a breeder seeks to develop traits such as size or intelligence by mating animals that present these characteristics—

and would probably not have come about naturally. But he stresses the fact that the variation observed in dog breeds must be inherent within its wild ancestors.[2]

Darwin's next theme is the "struggle for existence". Some individuals are better at competing for resources than others. Darwin reasons that the most successful will possess a variation of a particular trait or traits that gives them an advantage. These better-adapted individuals are likely to produce more offspring than others. Offspring inherit the favorable traits of their parents by means of biological inheritance. He could not describe the actual mechanism involved, because genes—the biological material that carries characteristics down the generations—had not yet been discovered; but he argues nonetheless that this process of natural selection results in a population with more individuals better adapted to their environment at a particular period. It is important to note that Darwin uses the phrase "survival of the fittest"* as a metaphor; what he was trying to say is that not all individuals will contribute equally to the next generation. The word "fittest" is not a value judgment: in biology the word designates the individuals that survive and, importantly, have offspring.

Darwin uses selective breeding to introduce the theme of gradualism,* showing how by small incremental steps, over successive generations, adaptations can arise. He admits that variations in domestic animals are not likely to be beneficial to the animals themselves, but are for "man's own use or fancy."[3] While these are not true adaptations, then, they do illustrate that organisms have the capacity to change gradually over time—illustrating both the concept of gradualism and that individuals can inherit traits from their parents.

Language and Expression

On the Origin of Species was written for a well-educated general reader skilled in critical thought. It is not necessary to have a scientific background to understand the text, written as it is in the standard prose of the period.

Darwin presents his case in favor of evolution, implicitly rejecting creationism* (the doctrine that the biblical account of the creation of the world and its animals is perfectly correct). He uses the term "descent with modification"* rather than "evolution". The latter came to be more widely used in the later part of the nineteenth century, appearing only in the sixth edition of 1872; the word "evolved" is only used once, in the work's very last sentence.[4]

Darwin uses the term "natural selection" to explain how organisms diversify and adapt. He makes only a few direct references to intelligent design*, and writes solely about the nature of organisms in terms of their physical properties and nature. He does his best to avoid stating explicitly how his theory undermines the biblical account of the origins of life. His argument was clear to his readers, though, and Darwin's contentious ideas would give rise to positivism*—which argues that information derived from sensory experience, and interpreted through reason and logic, forms the only source of authoritative knowledge.

Throughout the book, Darwin uses analogies, such as his comparison of domestic and wild animals; but it was not only his references to domesticated animals, but to domestic life, that enabled readers to identify with his writings easily.[5] One wonderful example of this is when he describes how he uses a teacup to scoop up mud from a local pond. In this way his ideas are expressed informally, and were accessible to the Victorian reader. He aimed to reach a wide audience: "I sometimes think that general and popular Treatises are almost as important for the progress of science as original work."[6] Darwin succeeded: *On the Origin of Species* is now a world-famous work.

1. Charles Darwin, *On the Origin of Species by Means of Natural Selection, or the Preservation of Favored Races in the Struggle for Life*. Introduction and Notes by Gillian Beer (1996, 2008), (Oxford: Oxford University Press, 1860).
2. Darwin, *Origin*.
3. Darwin, *Origin*.
4. Mark Ridley, *How to Read Darwin* (London: Granta Books, 2006).
5. George Levine, *Darwin the Writer* (Oxford: Oxford University Press, 2011).
6. Charles Darwin, in a letter to Thomas Huxley (1865), in *Darwin's Island: The Galapagos in the Garden of England* by Steve Jones (London: Little Brown, 2009).

MODULE 6
SECONDARY IDEAS

KEY POINTS

• Darwin's secondary ideas are sexual selection* (the process by which evolution occurs through the selection of a mate), gradualism* of species, and speciation* (the formation of distinct species).

• These secondary ideas, especially sexual selection and speciation, are still relevant to contemporary evolutionary* studies.

• The most important secondary idea is sexual selection, which is still an important factor in natural selection* as understood in modern evolutionary theory.

Other Ideas

A key subordinate idea in Charles Darwin's *On the Origin of Species by Means of Natural Selection* is the component of natural selection known as "sexual selection." This theory partly explains a paradox: how is it that some characteristics evolve when they reduce an individual's chances of survival—the peacock's tail, for example?

Another important secondary idea concerns the origin of new species—speciation*. Darwin's concept of evolution by natural selection requires "slow and gradual accumulation of numerous, slight yet profitable" modifications—termed gradualism.[1] In chapter 6 he highlights a problem, "Firstly, why, if species have descended from other ancestral species by insensibly fine gradations [over a long period of time] do we not everywhere see innumerable transitional forms?"[2] Addressing here the problem of the scarcity of transitional types in the fossil record, he goes on to discuss at great length how there must be some method in natural selection that leads to the generation of distinct species.

> "The sexual struggle is of two kinds: in the one it is between the individuals of the same sex, generally the males, in order to drive away or kill their rivals, the females remaining passive; while in the other, the struggle is likewise between the individuals of the same sex, in order to excite or charm those of the opposite sex, generally the females, which no longer remain passive, but select the more agreeable partners."
> ——Charles Darwin, *The Descent of Man, and Selection in Relation to Sex*

Exploring the Ideas

There are two forms of sexual selection. In intrasexual selection*, members of the same sex compete for opportunities to mate; red deer stags, for example, compete for access to females by clashing antlers. In intersexual selection*, members of the opposite sex are attracted to particular characteristics, such as extravagant and colorful tail feathers, as seen in male peacocks. If the peacock males chosen by the peacock hens are genetically different from those of their rivals, then natural selection occurs.[3]

According to Darwin, evolution by natural selection is a gradual process. New species are formed by slow incremental changes that take place over long periods of time. One of the major difficulties for Darwin was the absence in the fossil record of evidence of these changes, in the form of intermediary types.

Even today scientists are unable to find a fossil record for many intermediate forms. The evolutionary biologists* Stephen Jay Gould* and Niles Eldredge proposed an explanation for these gaps, suggesting that they are real, and represent periods of stability when species did not change much for millions of years. They believe these periods of stability are then followed by times of rapid change that result in new species, termed "punctuated equilibrium."[4]

According to this theory, changes resulting in a new species do not normally come from small, gradual shifts in the mainstream population, but occur instead in a small subset of the population, such as those living at the edge of the habitat or in a small isolated group. When the environmental conditions change, these "peripheral" or "geographic isolates" experience intense selection and speedy change due to both the altered environment and their small population size. They do not leave fossils reflecting the intermediate stages because their numbers are relatively few, and because of their isolated location. These new successful types can then spread out across the geographic area of the ancestral species.[5]

Indeed, the finches that Darwin collected on the Galapagos islands on his celebrated voyage on the *Beagle** each evolved from their ancestor on the mainland to fit a specific habitat on the remote islands in the Pacific. The theory of punctuated equilibrium does not imply that evolution *only* happens in rapid bursts. Scientists have also observed gradual evolution—notably in our own species.[6]

Does speciation imply that organisms exist as definite types? Darwin had reservations about the existence of distinct species: "Nor shall I here discuss the various definitions which have been given of the term species. No one definition has as yet satisfied all naturalists; yet every naturalist knows vaguely what he means when he speaks of a species."[7] In the second half of the twentieth century, over 20 different definitions of "species" were proposed.[8] Some argue that such a large number of definitions is itself proof that distinct species do not exist, and prefer to view living organisms as part of a gradually changing continuum rather than as distinct entities. Categories such as species, they say, are more a product of the human mind, which has a tendency to classify things for simplicity's sake, rather than being a true reflection of nature.[9]

Overlooked

Despite its importance to evolutionary theory, the concept of sexual selection was largely ignored for nearly a century, because people found it difficult to comprehend how animals could exhibit choice. Darwin himself had a problem with this notion, as animals were then regarded as robots at the mercy of their instincts. So, though scientists nowadays understand that choice as a mechanism can exist,[10] it was a difficult conceptual jump to make in Darwin's time.

Sexual selection has become central to modern evolutionary

biology and behavioral ecology* (a field in which animals' environments are used to explain their behavior). This is in part due to the British statistician Ronald Fisher*, who developed a model for sexual selection in 1915. He suggested that traits such as a male peacock's tail can evolve if there is a genetic basis for both the trait itself (the tail) and the sexual preference for the exaggerated version of that trait. If females carry a gene for the preference and their sons have a gene enabling them to develop the preferred trait, this results in a proportional increase in both the trait and the preference for it.[11] Fisher's work, however, was overlooked until the 1970s when it was discovered that female choice does indeed exert a very powerful effect on male traits.[12] In the 1970s and 1980s other models were also proposed for a review.[13]

1. Charles Darwin, *On the Origin of Species by Means of Natural Selection, or the Preservation of Favoured Races in the Struggle for Life* (London: John Murray, 1859), 103.

2. Darwin, *Origin*.

3. J. R. Krebs and A. Kacelnik, "Decision-making," in *Behavioural Ecology: An Evolutionary Approach* (Oxford: Blackwell Scientific, 1991), 105–36.

4. Niles Eldredge and S. J. Gould. "Punctuated equilibria: an alternative to phyletic gradualism", in *Models in Paleobiology*, ed. T. J. M. Schopf (San Francisco: Freeman Cooper, 1972), 82–115.

5. Gould, "Punctuated equilibria," 82–115.

6. W. A. Haviland and G.W. Crawford, *Human Evolution and Prehistory* (Cambridge, MA: Harvard University Press, 2002).

7. Darwin, *Origin*.

8. J. A. Mallet, "Species definition for the modern synthesis," *Trends in Ecology and Evolution* 10 (1995): 294–9.

9. Daniel Elstein, "Species as a Social Construction: Is Species Morally Relevant?" *Journal for Critical Animal Studies* 1, no. 1 (2003): 53–71. See also John Wilkins, *Species: A History of the Idea*

(Oakland: University of California Press, 2011).

10. Krebs and Kacelnik, "Decision-making," 105–36.

11. Malte B. Andersson, *Sexual Selection*: *Monographs in Behavior and Ecology* (Princeton: Princeton University Press, 1994).

12. Andersson, *Sexual Selection.*

13. Andersson, *Sexual Selection.*

MODULE 7
ACHIEVEMENT

KEY POINTS

* Darwin was the first to succeed in proposing a mechanism for explaining species change. It remains the overriding paradigm* (conceptual model) today.

* His work of over 21 years allowed him to present the strongest arguments for natural selection*. *On the Origin of Species* went on to influence many other disciplines, including the life sciences.

* Darwin's ignorance of the mechanisms of inheritance through the biological material now known as genes—a field of inquiry that was not established until 1906—meant that he misunderstood how traits are passed down the generations.

Assessing the Argument

Charles Darwin's *On the Origin of Species by Means of Natural Selection* argues one key point: that natural selection is the driving force behind evolution*. In the absence of clear evidence for evolution in the fossil record, Darwin made his case by considering shared ancestry,[1] pointing to the fact that animals living in close proximity to each other tend to share a closer ancestry.[2] He also introduced the concept of gradualism* as it applies to organisms.[3]

Knowing that he was courting controversy with the text's publication, Darwin devoted one chapter to answering all the questions he thought would arise, naming it "Difficulties on Theory."

At the time of the work's publication in 1859, Charles Lyell,* Thomas Henry Huxley*, and the botanist* Joseph Hooker* gave Darwin their professional and personal support. These were invaluable friends, who helped to establish Darwin's intellectual priority*—the acknowledgment that it was Darwin's work, and not anyone else's, that had made the greatest step forward, thereby making sure he received the credit for his major discovery.

This became critical in 1858 when it emerged that the naturalist* Alfred Russel Wallace* had also hit upon the same theory—natural selection—as the mechanism for species change. It was feared that Darwin might lose his intellectual priority to the younger man. Later, Lyell and Huxley would become a defense-and-attack team when *On the Origin of Species* became suddenly—and internationally—controversial. Darwin maintained his notable academic reputation and went on to publish three more major works on evolutionary theory, despite the fact that his theory of natural selection was widely criticized on certain points (the mechanism of inheritance was not yet understood; there were also issues regarding the age of the earth),[4] and the challenge it presented to orthodox religious Christian ideas.

"It will live as long as the 'Principia' of Newton ... Darwin has given the world a new science and his name should in my opinion stand above that of every philosopher."
——Alfred Russel Wallace, Wallace Letters Online

Achievement in Context

Darwin was certainly not the first person to raise the topic of evolution; it had been a topic ever since the Enlightenment*.[5] His own grandfather Erasmus Darwin* published several works on the matter, including his famous poem "The Origin of Society" (1803), on natural history and the relatedness of all life forms. Another celebrated precursor was the French biologist Jean-Baptiste Lamarck*, who in 1809 proposed that species could transform and were not, then, fixed entities. The power of Lamarck's writing was such that he made belief in evolution a respectable position to hold. Naturalists Étienne Geoffroy Saint-Hilaire* and Robert Grant* went on to become advocates of this idea, as did the Scottish writer Robert Chambers* who, in his *Vestiges of the Natural History of Creation* (1844), suggested that all complex life evolves from simpler forms. Naturally, the clergy opposed the book because it called traditional religious views into question, while scholars such as Adam Sedgwick* criticized it for its superficiality on scientific grounds.[6] Nonetheless, it remained extremely popular among the general public, partly because of its accessible style,[7] and it undoubtedly prepared Victorian society for the idea of evolution.[8]

The first print run of *On the Origin of Species* was of 4,200 copies, which sold out on its first day; five further editions were published during Darwin's lifetime. While Chambers' *Vestiges* continued to outsell it,[9] *On the Origin of Species* ultimately had the most impact and was considered more respectable, partly due to the efforts of Darwin's close friends, and partly because of Darwin's own position as a notable scientist.[10]

Limitations

Darwin's ideas have at times been applied to social policy. Darwin himself had already applied his theory of natural selection more broadly. In *The Descent of Man, and Selection in Relation to Sex* (1871) he wrote, "All ought to refrain from marriage who cannot avoid abject poverty for their children," referring to the struggle for existence. "Otherwise he would sink into indolence, and the more gifted men would not be more successful in the battle of life than the less gifted."[11]

Such statements gave rise to social Darwinism*—the theory that, as with other species, humans are subject to natural selection, with some races being more genetically advanced than others. In this approach, biological concepts, like the survival of the fittest*, are applied more broadly to economics, politics, and sociology. The idea that societies start out as "primitive" and advance toward civilization was first proposed by the English biologist Herbert Spencer*, who also developed the notion that certain (Western) cultures were superior.

Eugenics—a scientific method aimed at improving the qualities of humans through selective breeding—developed from this idea. It was Darwin's cousin, Francis Galton,* who first proposed the science of eugenics in 1883. He advocated the imposition of social controls including sterilization for those deemed "unfit." The idea that "unfit" individuals, such as those afflicted with physical or mental disabilities, should be stopped from reproducing was implemented in several countries, including Britain, Canada, Germany, Sweden, and the United States.[12]

The German philosopher Friedrich Nietzsche* warned that, taken out of context, the ideas in *On the Origin of Species* could be interpreted as nihilistic*—that is, suggesting that our existence is meaningless. This could absolve the individual of any moral or social responsibility, for if life has no purpose, why value it?

But the theory of evolution cannot be understood in terms of value judgments or what is *desirable* or *good*—a warning heeded as early as 1903 by the British philosopher George Edward Moore* who, in his book *Principia Ethica*, argues that it would be a mistake to try to define the concept *good* in terms of some natural property. This would be to commit what he termed the "naturalistic fallacy."*[13]

Others argue that if something has evolved, it must be good—even the social and economic system of capitalism,* and the pursuit of private profit that this entails. It has been pointed out, however, that "capitalism is justified on the basis that it is an expression of *the survival of the fittest* and the survial of the *good*. But by doing so we commit the naturalistic fallacy because good has been defined as something other than itself, rather as *the survival of the fittest.*"[14]

In conclusion, it is dangerous to base human morality on the laws and concepts that emerge from the natural sciences.

1. Charles Darwin, *On the Origin of Species by Means of Natural Selection, or the Preservation of Favored Races in the Struggle for Life.* Introduction and Notes by Gillian Beer (1996, 2008), (Oxford: Oxford University Press, 1860).

2. Darwin, *Origin*.

3. Mark Ridley, *How to Read Darwin* (London: Granta Books, 2006).

4. Ridley, *How to Read Darwin*.

5. Edward J. Larson, *Evolution: The Remarkable History of a Scientific Theory* (New York: Modern Library, 2004).

6. Adam Sedgwick, "Review of *Vestiges*," *Edinburgh Review* 82 (July 1845): 1–85.

7. James A. Secord, *Victorian Sensation: The Extraordinary Publication, Reception, and Secret Authorship of Vestiges of the Natural History of Creation* (Chicago: University of Chicago Press, 2000).

8. Lois N. Magner, *A History of the Life Sciences* (New York; Basel: Marcel Dekker, 1994), 257–316.

9. Magner, *A History of the Life Sciences*.

10. George Levine, *Darwin the Writer* (Oxford: Oxford University Press, 2011).

11. Charles Darwin, *The Descent of Man, and Selection in Relation to Sex* (London: John Murray, 1871).

12. Dennis Sewell, *The Political Gene: How Darwin's Ideas Changed Politics* (London: Picador, 2009).

13. G. E. Moore, *Principia Ethica* (Cambridge: Cambridge University Press, 1993).

14. Julia Tanner, "The Naturalistic Fallacy," *Richmond Journal of Philosophy* 13 (2006): 1–6.

MODULE 8
PLACE IN THE AUTHOR'S WORK

KEY POINTS

* Darwin's life's work was to explain the natural world through an understanding of evolution*.

* *On the Origin of Species* was the foundation on which all Darwin's later theories were built.

* *On the Origin of Species* and *Descent of Man* together made Darwin's reputation as the founder of evolutionary theory; his name has become synonymous with the principle of evolution itself.

Positioning

From an early age, Charles Darwin had an inquisitive nature and found the natural world around him fascinating. It is clear from reading *On the Origin of Species by Means of Natural Selection* that he found the biblical explanation for the creation of this natural world unsatisfactory. In his early twenties, he read widely on various subjects, many of which argued in favor of the active role of the divine in the affairs of humankind; natural theology*—the idea that the beauty and complexity of nature is itself evidence for the existence of God—was the dominant scientific position at that time.

Darwin regarded these explanations as unreasoned; throughout his life, his purpose was to explain the natural world he observed around him. The theories presented in *On the Origin of Species* and his subsequent books were thoughtful and detailed accounts of his ideas.

The central questions found in Darwin's books took shape early on in his life, perhaps soon after his return from the voyage aboard HMS *Beagle**.[1] In 1838, he read the Scottish anatomist* and philosopher Charles Bell's* *Essays On The Anatomy And Philosophy Of Expression* (1824), which concluded that man's emotions are unique.[2] From the notes in the margins of Darwin's original copy of Bell's book, it is clear that he did not agree,[3] and over the course of 30 years he gathered evidence that would eventually appear in *The Expression of the Emotions in Man and Animals* (1872). In this book Darwin uses descent from a common ancestor (evolution) to explain similarities between the emotions of humans and other animals.

> *"As long as man and all other animals are viewed as independent creations, an effectual stop is put to our natural desire to investigate as far as possible the causes of Expression ... The community of certain expressions in distinct though allied species ... is rendered somewhat more intelligible, if we believe in their descent from a common progenitor. He who admits on general grounds that the structure and habits of all animals have been gradually evolved, will look at the whole subject of Expression in a new and interesting light."*
> ——Charles Darwin, *The Expression of the Emotions in Man and Animals*

Integration

Evolution is the overarching theme in all Darwin's major texts: *On the Origin of Species* (1859), *The Variation of Animals and Plants*

Under Domestication (1868), *The Descent of Man* (1871), and *The Expression of the Emotions in Man and Animals* (1872). In each book, Darwin explains how natural selection* has shaped humans and other animals we see today.

In the two-volume *Variation of Animals and Plants*, Darwin expands his discussion about inheritance and variation in plants and animals. He introduces his theory of pangenesis*—the idea that small pieces of information in the form of gemmules (hypothetical particles) float from all parts of the adult body and go on to make the information that forms the offspring. In the very last paragraph of this book he argues more directly than in *On the Origin of Species* against the belief in evolution by design: "An omniscient Creator must have foreseen every consequence which results from the laws imposed by Him ... Can it with any greater probability be maintained that He specially ordained for the sake of the breeder each of the innumerable variations in our domestic animals and plants;—many of these variations being no service to man, and not beneficial, far more often injurious, to the critters themselves? Did He ordain that the crop and tail-feathers of the pigeon should vary in order that the fancier might make his grotesque pouter and fantail breeds? ... However much we may wish it, we can hardly follow ... that variation has been led along certain beneficial lines."[4]

The Descent of Man expands on the ideas in *On the Origin of Species*, introducing further evidence for evolution in plants and animals. Darwin also sets out new details about human origins, focusing on the evolution of human mental and moral faculties. He argues that the difference between the mental faculties of humans and those of other animals is one of degree rather than kind. He also expands on his theory of sexual selection*—a form of natural selection in which the selection of a mating partner plays the principal role—that had previously figured in *On the Origin of Species*.[5] *The Descent of Man* was considered equally thought-provoking and scandalous.

Further evidence for human evolution from an ape-like ancestor appears in Darwin's subsequent book *The Expression of the Emotions in Man and Animals*: "With mankind some expressions, such as the bristling of the hair under the influence of extreme terror, or the uncovering of the teeth under that of furious rage, can hardly be understood, except on the belief that man once existed in a much lower animal-like condition."[6]

Clearly, Darwin spent much of his life arguing in favor of evolution as a framework for explaining adaptations in humans and other animals. His books, and the influential way in which he considered human beings to be animals, revolutionized thinking in both the sciences and the humanities.

Significance

The debate about evolution in the 1850s was heated. When Darwin published in 1859, he transformed this debate largely because he was able to take all the current information and forge it into a coherent argument.

Darwin's first converts were his close friends Thomas Henry Huxley* and Joseph Dalton Hooker*, who then became vocal advocates for evolution. Both men were academic free thinkers who succeeded in their campaign to remove the influence of religious dogma from science. They backed the liberal* Anglican movement, which accepted evolution and opposed the traditionalists who had rejected it.[7]

Darwin's influence soon stretched beyond academic circles. When the role of archbishop of Canterbury became vacant in 1896, it was filled by Frederick Temple*, an affirmed supporter of evolution.[8] When Darwin died, he was buried in London's Westminster Abbey near the great scientist Isaac Newton*, a mark of great public prestige, and both social and religious acceptance.[9]

Today the theory of natural selection is the unifying concept of the life sciences.

---

1. Mark Ridley, *How to Read Darwin* (London: Granta Books, 2006), 8–15.
2. Charles Bell, *Essays on the Anatomy and Philosophy of Expression* (Montana: Kessinger Publishing, 2008).
3. Ridley, *How to Read Darwin*.
4. Charles Darwin, *The Variation of Animals and Plants Under Domestication* (London: John Murray, 1868).
5. Charles Darwin, *The Descent of Man, and Selection in Relation to Sex* (London: John Murray, 1871).
6. Charles Darwin, *The Expression of the Emotions in Man and Animals* (London: John Murray, 1872).
7. Edward J. Larson, *Evolution: The Remarkable History of a Scientific Theory* (New York: Modern Library, 2004), 79–111.
8. Philip Kitcher, *Living with Darwin: Evolution, Design, and the Future of Faith* (New York; Oxford: Oxford University Press, 2007).
9. Janet Browne, *Charles Darwin: A Biography. Vol. 2: The Power of Place* (London: Pimlico, 2003), 5.

SECTION 3
IMPACT

THE FIRST RESPONSES

KEY POINTS

* On publication, *On the Origin of Species* was criticized because species change opposed the prevailing belief that organisms are fixed, and because it did not mention a divine Creator.

* The scientific mainstream accepted that species change exists; Darwin modified subsequent editions to include references to a Creator.

* The anatomist Thomas Huxley* undertook a spirited public defense of *On the Origin of Species*, as did religious friends such as the clergyman Charles Kingsley*.

Criticism

Charles Darwin's *On the Origin of Species by Means of Natural Selection* and his theory of evolution* provoked widespread criticism, principally focusing on his omission of an intelligent Creator—God. The concept of changing species was considered blasphemous, too, as it implied that God had created beings that were not perfect. It was not just the Church that disapproved. Darwin's close friend and mentor, the geologist* Adam Sedgwick*, was disappointed in his pupil and feared the moral implications of the theory.[1]

Professional envy also played a part. In an attempt to discredit the theory, the anatomist* Richard Owen* anonymously penned a long and malicious article in the *Edinburgh Review* in April 1860. Owen stated that Darwin's proposed mechanism of natural selection* was wrong, arguing for "the continuous operation of the ordained becoming of living things."[2]

There was also more general criticism on scientific grounds, some of which Darwin took seriously enough to address and refute in later editions. The first came from the zoologist* St. George Jackson Mivart, who argued that natural selection did not explain the early stages in the development of an organ such as the eye.[3] If the eye came about through small incremental steps, he argued, then there must have been a stage when it had no useful function and so gave no selective advantage.

Another objection came from the Scottish engineer Fleeming Jenkin* in 1867. He suggested that in a large population, an adaptive trait existing among a few individuals would soon disappear when they interbred with others who lacked the feature. Jenkin supposed that genetic factors would divide during interbreeding and that a favorable trait would be diluted in subsequent generations.[4]

"Science demonstrates incessant past changes, and dimly points to yet earlier links in a more vast series of development of material existence; but the idea of a beginning, or of creation, in the sense of the original operation of the divine volition to constitute nature and matter, is beyond the province of physical philosophy."

—— Rev. Baden Powell, *Philosophy of Creation*

Responses

Darwin addressed the theological criticism in the second edition of *On the Origin of Species*, published just two months after the first in 1860, adding a few sentences on the Creator. He also quoted support for his idea of natural selection by a "celebrated cleric"—Reverend Charles Kingsley—whom he did not name but who, he wrote, "has gradually learnt to see that it is just as noble a conception of the Deity to believe that He created a few original forms capable of self-development into other and needful forms, as to believe that He required a fresh act of creation to supply the voids caused by the action of His laws."[5]

In response to Mivart's critique, Darwin argued that an organ could be beneficial during its early stages of development. For example, the eye would have started out as a light-sensitive eyespot. Over time, other developments would have occurred, granting greater benefit, and so through small steps, the eye evolved to become the complex organ it is today. Now, we understand that Darwin's explanation is correct: "a fortuitous novelty may confer a subtle advantage." During his lifetime, though, this issue remained a problem and one of the factors that meant natural selection was doubted as a theory to explain evolution.[6]

Darwin was also unable to show evidence for a mechanism of inheritance. In fact, less than a year after Fleeming Jenkin raised this question, Gregor Mendel*, the founder of genetics*, published a paper that would have answered his objections. Unfortunately, Mendel's work did not gain acceptance until its rediscovery in 1900. In his 1866 article, Mendel showed the action of "invisible factors" (or genes*), in coding visible traits. He showed that these "factors" were indivisible and did not, therefore, blend during interbreeding as Jenkin had presumed.[7]

Following the objections, Darwin decided it was necessary to add some supplementary processes to the theory of natural selection in order to show that evolution could take place at a faster pace. In the sixth and final edition of *On the Origin of Species*, he writes that natural selection "aided in an important manner, by inherited effects of the use and disuse of parts; and in an unimportant manner, that is in relation to adaptive structures, whether past or present, by the direct action of external conditions."[8]

Here, Darwin reverts to the inheritance of acquired characteristics as first proposed by the French biologist Jean-Baptiste Lamarck.* He also proposes a new theory,"pangenesis,"* according to which the sex cells (the sperm and ovum) absorb a set of particles, which represent all the organs and tissues of the adult body. In this way, characteristics acquired during the adult stage of life are passed on to offspring. This theory is similar to the one set out by the ancient Greek philosopher Democritus.*[9]

Conflict and Consensus

While *On the Origin of Species* was successful in convincing the scientific community to accept evolution, Darwin was not able to satisfy everyone that natural selection was the primary mechanism for species change, largely thanks to Jenkin's criticism. And so after 1870, natural selection fell out of favor and Lamarck's theory of transmutation*—the idea that if an organism adapts its body to its environment during the course of its life, those changes are passed on to its offspring—became widely accepted again.[10]

Other theories also sprang up at this time, such as orthogenesis,* now an obsolete biological hypothesis. Popularized by Theodor Eimer's* *Organic Evolution as the Result of the Inheritance of Acquired Characteristics According to the Laws of Organic Growth* (1890), orthogenesis suggests that organisms have an inborn tendency to evolve in a consistent and steady way, because of some internal mechanism or driving force.[11]

1. R. Weikart, *From Darwin to Hitler: Evolutionary Ethics, Eugenics and Racism in Germany* (London: Palgrave Macmillan, 2004).

2. Mark Ridley, *How to Read Darwin* (London: Granta Books, 2006), 8–15.

3. St. George Jackson Mivart, *On the Genesis of Species* (Cambridge: Cambridge University Press, 2009).

4. Fleeming Jenkin, "Review of *The Origin of Species*," *The North British Review* 92, no. 46 (June 1867): 277–318.

5. Charles Darwin, *On the Origin of Species by Means of Natural Selection, or the Preservation of Favoured Races in the Struggle for Life*, 2nd edition (London: John Murray, 1860), 481.

6. Mario Livio, *Brilliant Blunders: From Darwin to Einstein: Colossal Mistakes by Great Scientists That Changed Our Understanding of Life and the Universe* (New York: Simon & Schuster, 2013).

7. Gregor Mendel, "Versuche über Pflanzenhybriden," *Verhandlungen des naturforschenden Vereins Brünn* (1866).

8. Charles Darwin, *On the Origin of Species by Means of Natural Selection, or the Preservation of Favoured Races in the Struggle for Life*, 6th edition (London: John Murray, 1872).

9. Edward J. Larson, *Evolution: The Remarkable History of a Scientific Theory* (New York: Modern Library, 2004).

10. Larson, *Evolution*.

11. Theodor Eimer, *Organic Evolution as the Result of the Inheritance of Acquired Characteristics According to the Laws of Organic Growth* (London: Macmillan, 1890).

MODULE 10
THE EVOLVING DEBATE

KEY POINTS

- *On the Origin of Species* has been deeply influential, notably in our understanding of human behavior—is it hardwired or socially constructed?—and the concept of human "uniqueness."

- In Darwinian-influenced biological anthropology*, humans are put in an animal context with other closely related species, and our behavior and variation are studied as they are in other animals.

- Darwin's work finally permitted scholars to make the necessary link between humans and our fellow primates; humans are now classified as animals and grouped in the family *Hominidae*, or the great apes.

Uses and Problems

On the Origin of Species by Means of Natural Selection by Charles Darwin proposed that species evolve by natural selection*, but by 1870 Darwin had serious doubts as to whether this mechanism could fully explain evolution*. Fervent supporters of the theory, the scientists Alfred Russel Wallace* and the German naturalist August Weismann,* however, continued his work, founding a movement known as neo-Darwinism*.

Weismann scrutinized the concept that parents are able to pass on particular qualities (such as ability on the piano or a brawny body) to their offspring, going so far as to conduct experiments to see if scars on a parent would be passed on to the child. In a paper written in 1883, Weismann declared that acquired characteristics could not, in fact, be inherited. Sex cells (the collective name for sperm and egg) are segregated at an early stage in their development, so cannot be affected by any changes taking place in the parent's body once that has occurred.[1] By 1885, he had identified the nucleus of germ cells as the carriers of genetic* information. Germ cells are biological cells that give rise to the gametes—the cells that develop from the fusion of sperm and egg—of any organism that reproduces sexually.

In 1889, in defense of natural selection, Wallace wrote *Darwinism*, a book that presented his own ideas on speciation* (the process by which species are formed) and highlighted the importance of environmental pressures in forcing species to adapt to local habitats. When populations are isolated geographically and therefore have a smaller choice of mates, they begin to diverge until there are two separate species.[2] This theory came to be known as "the Wallace effect."[3] Current research supports this idea.[4,5]

In the early 1900s, the botanists* Hugo de Vries* and Carl Correns* rediscovered the work of Gregor Mendel*, the founder of genetics. Mendel's work showed that offspring retain distinct characteristics from each parent rather then a mixture of traits blended together. However, misunderstandings about Mendel's ideas led scientists to think that new characteristics or even species would suddenly appear.[6] DeVries went on to suggest that new species arise by a process of mutation*—abrupt changes in inheritable characteristics—and not by natural selection. In contrast to Darwin's idea of gradual change, deVries thought that species evolve in sudden, dramatic changes. He based this "theory of mutation"* on his work with the evening primrose plant. He saw that the original plant sometimes had offspring with significant visible differences, in, for example, leaf shape or plant height. De Vries then designated these as new species.

Essentially, de Vries had the right idea but for the wrong reasons; most of the variants he observed were due to aberrant chromosomal segregations (when paired chromosomes split and migrate to opposite ends of the nucleus), and were not mutations at all.

For several decades (1900–30), there were two opposing schools of thought. One upheld the Darwinian view of evolution by natural selection; the other believed that evolution was the result of a series of drastic mutations, as de Vries had proposed. In an attempt to confirm the latter theory, the US anthropologist* Thomas Hunt Morgan* began working on the common fruit fly, and managed to establish a link between Mendel's work on the common pea plant and Walter Sutton's* work that identified genes* as the carriers of hereditary information. Morgan's experiments showed that mutations did not suddenly create new species, but increased variations within a population.[7] The details of his work are published in *The Mechanism of Mendelian Heredity* (1915).[8]

From 1930 to 1950, scientists began to draw together the ideas of Mendel and Darwin. Darwin's theory of natural selection began to be more widely accepted, supported by new research from the fields of genetics and population statistics. This became known as the new synthesis period* and gave rise to the field of evolutionary biology*—inquiry into the mechanisms of inheritance conducted in the light of evolutionary theory.

"If then, said I, the question is put to me would I rather have a miserable ape for a grandfather or a man highly endowed by nature and possessing great means and influence and yet who employs those faculties for the mere purpose of introducing ridicule into a grave scientific discussion—I unhesitatingly affirm my preference for the ape."

—— Thomas Henry Huxley, "Letter to Dr. Dyster,"
September 9, 1860

Schools of Thought

Two major schools of thought emerged during the new synthesis period: modern evolutionary theory and sociobiology* (the field of science that suggests that human behavior has resulted from evolution). The Ukrainian evolutionary biologist Theodosius Dobzhansky* was a prominent figure in evolutionary theory. His *Genetics and the Origin of Species* (1937) combined Darwin's theory of natural selection together with Mendel's genetics and research from biological disciplines.[9]

In sociobiology, social behavior is explained by the notion that it has evolved to produce a beneficial outcome for the individual. The English scientist W. D. Hamilton's* work on kin selection* during the 1960s helped develop this new discipline.[10] Hamilton showed that closely related kin display more altruistic (selfless) behavior toward each other—a fact that is now known as Hamilton's rule*. Hamilton explained how altruistic behavior such as eusociality* in insects (that is, the existence of sterile worker classes) could have developed from kin selection.[11]

In Current Scholarship

The 1980s and 1990s saw a revival of structuralist* ideas in evolutionary biology; structuralism is an intellectual current, influential in fields such as anthropology* and the study of language, in which (very roughly) different components of a language or a culture, say, are considered to be part of a system. This was partly due to the work of the biologists Brian Goodwin* and Stuart Kauffman*, who emphasized the contribution of self-organization to the course of evolution.[12] (Self-organization is when the components of a system interact to become ordered, in a previously disorganized system.)

From the 1980s onward, new data began to accumulate, leading scientists to understand that it is not different sets of proteins that control the way animals develop their phenotype* (visible characteristics). Instead, it is changes in the distribution of a small set of proteins that are common to all animals.[13] These proteins are called the *developmental-genetic toolkit*.[14] This knowledge had a great impact on the discipline of phylogenetics* (the study of the evolutionary relationships between a group of species), paleontology* (the study of fossilized remains) and comparative developmental biology,* and gave rise to a new discipline—evolutionary developmental biology.[15]

Modern biologists are less concerned than Darwin had been with whether natural selection explains adaptation, because only 5 percent of genetic change is adaptive. As a consequence of the discovery of DNA*—something, of course, Darwin knew nothing about—they place more emphasis on random evolutionary change.

(DNA is the material, carried in genes, that provides instructions for the growth and functioning of all living organisms.)

It is now accepted that two main processes cause evolutionary change: natural selection and random genetic drift* (the process by which genetic information changes over time as animals reproduce). Evolution is not only driven by natural selection, as Darwin argued. It can also happen by chance if there are two equally good versions of a gene or allele* (gene variant) and one is luckier than the other, over generations, in spreading through a population.

To give a simple illustration, a child inherits the gene code for black hair from its mother and a second gene code for red hair from its father. This individual then has a child. It can only pass on one copy of the gene and, by chance, passes on to its child the gene that codes for black hair. Over generations, if this scenario is replayed, the black hair color becomes dominant in the population and the code for red hair gets lost. Darwin, however, only knew about the observable form of organisms, and was more concerned with the evolution of these features.[16]

---

1. Friedrich Leopold August Weismann, *Die Entstehung der Sexualzellen bei den Hydromedusen: Zugleich ein Beitrag zur Kenntniss des Baues und der Lebenserscheinungen dieser Gruppe* (Jena: Fischer, 1883).

2. Alfred Russel Wallace, *Darwinism: An Exposition of the Theory of Natural Selection, with Some of Its Applications* (London: Macmillan & Co, 1889).

3. Edward J. Larson, *Evolution: The Remarkable History of a Scientific Theory* (New York: Modern Library, 2004).

4. Michel Durinx and Tom J. M. Van Dooren, "Assortative Mate Choice and Dominance Modification:

Alternative Ways of Removing Heterozygote Disadvantage," *Evolution* 63, no.2 (2009): 334–52.

5. J. Ollerton, "Speciation: Flowering time and the Wallace Effect," *Heredity* 95, no. 3 (2005): 181–2.

6. Larson, *Evolution*.

7. Larson, *Evolution*.

8. Thomas Hunt Morgan, et al., *The Mechanism of Mendelian Heredity* (New York: Henry Holt and Company, 1915).

9. Larson, *Evolution*.

10. Joel L. Sachs, "Cooperation within and among species," *Journal of Evolutionary Biology* 19, no.5 (2006): 1415–8.

11. Martin A. Nowak, "Five rules for the evolution of cooperation," *Science* 314 (2006): 1560–3.

12. Peter Corning, *Holistic Darwinism: Synergy, Cybernetics, and the Bioeconomics of Evolution* (Chicago: University of Chicago Press, 2010), 95–99.

13. John R. True and Sean B. Carroll, "Gene Co-option in Physiological and Morphological Evolution," *Annual Review of Cell and Developmental Biology* 18 (2002): 53–80.

14. Cristian Cañestro, Hayato Yokoi, and John H. Postlethwait, "Evolutionary Developmental Biology and Genomics," *Nature Reviews Genetics* 8, no. 12 (2007): 932–42.

15. Jaume Baguñà and Jordi Garcia-Fernàndez, "Evo-Devo: the Long and Winding Road," *International Journal of Developmental Biology* 47, nos. 7–8 (2003): 705–13.

16. Mark Ridley, *How to Read Darwin* (London: Granta Books, 2006).

MODULE 11
IMPACT AND INFLUENCE TODAY

KEY POINTS

* *On the Origin of Species* remains as relevant today as it was upon publication.

* The debates between hardwired behavior (what Darwin termed "instinct") versus social constructivism* (the idea that behavior is often constructed by surrounding culture) still flourish, as does the question of how unique humans are among animals.

* Fervent opposition to the theory of evolution* still exists among some religious groups, who fear it leads to atheism.

Position

Inspired by Charles Darwin's theory of natural selection* as set out in *On the Origin of Species by Means of Natural Selection*, a new discipline—sociobiology*—arose that attempted to apply an evolutionary framework to social conditions.

It has been argued that some human behaviors are programmed in our genes* from birth—a point of view termed "biological determinism*." Sociobiological theories have been spearheaded by biologists such as Richard Dawkins* and Edward Osborne Wilson,* among others. In *The Selfish Gene* (1976), Dawkins proposes that we cannot escape what is dictated by our genes, no matter how hard we try.[1] Expanding this idea and applying it more broadly to society, Wilson has said that all societies "no matter how egalitarian, would always give a disproportionate share of power to men because of the fixed genetic differences between men and women."[2]

Evolutionary biologists* such as Stephen Jay Gould* and Richard Lewontin* have criticized this approach as a reductionist* interpretation (an oversimplification) of human behavior. Recent sociological and scientific studies have shown that the determination of gender differences are trends, not defined behaviors.[3,4] Hard-line determinists*, on the other hand, such as the evolutionary psychologist* Steven Pinker,* define behavior as "male" or "female." Speaking at the Tanner Lectures in 1982, Lewontin offered an insight into why determinist ideas have nevertheless proliferated: "What makes these various inequalities between individuals, races, nations, and the sexes so problematic for us is the obvious contradiction between the fact of inequality and the ideology of equality on which our society is supposedly built."[5]

So science has now replaced religion as the ultimate intellectual authority. In the past, an unequal society was justified as being the way God meant it to be . Some scientists now attempt to justify social inequalities using arguments from biology, claiming that inequality such as that between men and women is due to the inherent differences in the biology of the brain. These scientists are criticized for reinforcing dominant ideologies by using science to answer questions of a personal, social, or political nature.

"You could give Aristotle a tutorial. And you could thrill him to the core of his being. Yet not only can you know more than him about the world. You also can have a deeper understanding of how everything works. Such is the privilege of living after Newton, Darwin, Einstein, Planck, and their colleagues. I'm not saying you're more intelligent than Aristotle, or wiser. For all I know, Aristotle's the cleverest person who ever lived. That's not the point. The point is only that science is cumulative, and we live later."

—— Richard Dawkins, "Science, Delusion and the Appetite for Wonder"

Interaction

Natural selection suggests that human beings are the product of a long evolutionary process, as they are specialized animals among many others. In the nineteenth century, this view was controversial because it challenged the biblical doctrine of exceptionalism, which claims that humans are the pinnacle of God's creation. The idea created an identity crisis in Victorian society because many people refused to believe they could be so closely related to animals.[6]

Today, the issue of human uniqueness is debated just as widely as it was when the work was published in 1859.[7] The anthropologist Kim Hill* argues that what sets us apart from other animals is our dependence on culture and cooperation. The evolutionary psychologists Josep Call* and Michael Tomasello* have shown that great apes are able to determine the intentions of others and, in the unique case of humans, the ability to *share* intentions.[8] Tomasello concludes that this demonstrates the cognitive chasm between us and the other apes.[9]

However, "the difference in mind between man and the higher animals, great as it is, certainly is one of degree and not of kind."[10] This, undoubtedly, is still the dominant view opposing claims of human exceptionalism. Theologians* such as Mark Harris argue that scientific evidence used to back up claims of human uniqueness are only examples of the qualitative differences between ourselves and the other animals, because there is no one trait that sets humans apart from other primates.[11] Research into human origins has revealed more similarities with our biological cousins than it has differences.

The Continuing Debate

Conversely, the idea of evolution has been uncontroversial in mainstream science for almost a hundred years. Since the 1930s and 1940s, when modern evolutionary theory incorporated genetic* science into Darwinian theory, most denials of the idea of evolution have originated from religious groups who maintain their belief in the creation myth. Arguments against the theory of evolution include objections to evidence, scientific methods, morality, and plausibility.[12]

A central text of the intelligent design* movement is the book *Darwin on Trial* (1991) written by the lawyer Philip Johnson*. He uses a legal framework to structure his argument. To illustrate, the legal term "beyond a shadow of a doubt" (used in the rare cases when it is deemed there is sufficient evidence to say that something is true) is used to destroy a scientific theory.[13] However, the evolutionary biologist* Stephen Jay Gould* points out that a legal argument such as this is simply inappropriate when applied to science, because "science is not a discipline that claims to establish certainty."[14]

Certainly, the belief that evolution promotes atheism has produced vehement opposition to it.[15] Creationists claim that proponents of evolutionary theory are atheists who cannot see beyond material causes and facts, to the detriment of a complete understanding of the nature of existence.[16] But such claims are tenuous; a poll conducted in 2014 found that 40 percent of scientists in the United States believe in a god—similar to results conducted among the general American public.[17,18]

Religious literalists support the creation myth unswervingly, but there is evidence to suggest that the biblical book of Genesis is a retelling of earlier narratives—encompassing themes from the ancient Middle Eastern myth of Gilgamesh* and ancient Mesopotamian* understandings of the world. One Mesopotamian myth describes how the goddess Ninti was sent to heal another god called Enki, whose body had been ravaged by disease. The Sumerian word *Ninti* has a double meaning—both "rib" and "life"—a theme repeated in Genesis, in which Eve is created from Adam's rib.[19]

Indeed, the Christian religious scholars Bruce Waltke* and Conrad Hyers* caution against a literal interpretation of the creation myth on the grounds that it is aligned with these tenets of Mesopotamian science and religion, arguing against such a reading precisely because it leads to the denial of evolution. For them, current scientific knowledge should be incorporated into the creation narrative instead.[20] The Roman Catholic Church has done precisely that, reconciling its belief in a deity with evolution by advocating theistic evolution* (the idea that evolution is a process put into action and guided by the hand of God).

---

1. Richard Dawkins, *The Selfish Gene* (Oxford: Oxford University Press, 1990).

2. E.O. Wilson, "Human Decency Is Animal," *New York Times Magazine*, October 12, 1975.

3. B. J. Carothers and H. T. Reis, "Men and Women Are From Earth: Examining the Latent Structure of Gender," *Journal of Personality and Social Psychology* 104, no. 2 (2013): 385–407.

4. Daphna Joel et al., "Sex Beyond the Genitalia: The Human Brain Mosaic," *PNAS* 112, no. 50 (2015): 15468–73.

5. R. C. Lewontin, "Biological Determinism," The Tanner Lectures on Human Values, University of Utah, March 31, April 1 1982. Accessed February 5, 2016, http: //tannerlectures.utah.edu/_documents/a-to-z/l/lewontin83.pdf.

6. Edward J. Larson, *Evolution: The Remarkable History of a Scientific Theory* (New York: Modern Library, 2004).

7. H. Guldberg, "Restating the Case for Human Uniqueness," *Psychology Today*, November 8, 2010.

8. J. Bräuer, J. Call, and M. Tomasello, "Chimpanzees really know what others can see in a competitive situation," *Animal Cognition* 10 (2007): 439–48.

9. M. Tomasello et al., "Understanding and sharing intentions: the origins of cultural cognition," *Behavioral and Brain Sciences* 28, no. 5 (2005): 675–735.

10. Charles Darwin, *The Descent of Man, and Selection in Relation to Sex* (London: John Murray, 1872), 82.

11. Mark Harris, "Human uniqueness, and are humans the pinnacle of evolution?" *Science and Religion @ Edinburgh*, September 7, 2014, accessed February 16, 2016, http: //www.blogs.hss.ed.ac.uk/science-and-religion/2014/09/07/ human-uniqueness-and-are-humans-the-pinnacle-of-evolution/.

12. Peter Cook, *Evolution Versus Intelligent Design: Why All the Fuss? The Arguments for Both Sides* (Australia: New Holland Publishing, 2007).

13. Phillip E. Johnson, *Darwin on Trial* (Downers Grove, IL: InterVarsity Press, 1991).

14. S. J. Gould, "Impeaching a Self-Appointed Judge," *Scientific American* 267, no. 1 (1992): 118–21.

15. Lee Strobel, *The Case for a Creator: A Journalist Investigates Scientific Evidence That Points Toward God* (Grand Rapids, MI: Zondervan, 2004).

16. Phillip E. Johnson, "The Church of Darwin," *Wall Street Journal,* August 16, 1999.

17. Larry Witham, "Many Scientists See God's Hand in Evolution," *Reports of the National Center for Science Education* 17, no. 6 (November-December 1997): 33.

18. Bruce A. Robinson, "Beliefs of the U.S. Public about Evolution and Creation," accessed 14 April 2015, www.ReligiousTolerance.org.

19. Samuel Henry Hooke. *Middle Eastern Mythology* (Dover Publications, 2013), 115.

20. Conrad Hyers, *The Meaning of Creation: Genesis and Modern Science* (Louisville: Westminster John Knox, 1984).

MODULE 12
WHERE NEXT?

KEY POINTS

* The theories found in *On the Origin of Species* have never been disproven and will inform scientific investigations far into the future.

* Today, emerging research into DNA* and genetics* is dependent on the theory of natural selection*, and will probably continue to be so in the future.

* The theories put forward in *On the Origin of Species* form the basis for all contemporary biological sciences.

Potential

Charles Darwin's *On the Origin of Species by Means of Natural Selection* discusses the evolutionary* past of life on earth—but many people wonder what the evolutionary future may be for humans.

The Welsh geneticist Steve Jones* argues that human evolution is slowing down, because we are no longer subject to natural selection—the *fittest* individuals, that is, no longer drive evolutionary change. In the nineteenth century, when Darwin published *On the Origin of Species*, fewer than half of British children survived to 35 years of age. Today that number is around 95 percent,[1] largely due to the advances made in medicine in the treatment of various illnesses and diseases. Now those individuals deemed weakest continue to survive and have children. In Jones's words, "Darwin's machine has lost its power."[2]

The evolutionary psychologist* Geoffrey Miller* disagrees.

Viral and bacterial pathogens can now, due to modern technology such as airplanes, spread more easily to various parts of the globe. Miller predicts that epidemics will become important in shaping the human immune system, and will result in future generations of humans possessing stronger immune systems.

Although *On the Origin of Species* is no longer relevant in many regards, the ideas it proposes still dominate, and extend beyond the sciences to touch most people alive today.

> *"It will be possible, through the detailed determination of amino-acid sequences of hemoglobin molecules and of other molecules too, to obtain much information about the course of the evolutionary process, and to illuminate the question of the origin of species."*
> —— Linus Pauling, *Molecular Disease and Evolution*

Future Directions

The ancient Greek philosopher Aristotle* envisaged life as a Great Chain of Being*, with every organism occupying a place in a hierarchy, and there is still a "tendency in evolutionary discourse to describe life's history as a progression towards increasing complexity."[3] But there are cases where simple organisms—such as gutless tapeworms or blind cave fish—have arisen from more complex ones. This is termed "reductive evolution".

It has also been found that organisms can occasionally lose their ability to perform a function previously thought necessary to their survival, without it affecting their ability to survive or multiply. This was first discovered in the ocean-dwelling plankton *Prochlorococcus*—a common photosynthetic microorganism (an organism that can convert sunlight into energy). Researchers found that it had lost the gene* that helps to neutralize hydrogen peroxide, a compound that can destroy cells.[4] Instead, it relies on other nearby microorganisms to eliminate hydrogen peroxide from its environment.

For microorganisms, carrying genes and manufacturing proteins requires a great deal of energy, so discarding certain genes allows them to live more efficiently. The researchers who first made this observation have termed it the Black Queen hypothesis*—the principle that natural selection drives microorganisms to lose essential functions when there is another species nearby to perform them.[5] Evolution can favor coexistence between helpers that perform the function and beneficiaries that require it. The fitness of helpers is not reduced, resulting in a stable coexistence that can lead to the evolution of mutualism,* where two organisms of different species cooperate and benefit from the arrangement. These helpers are not, therefore, social altruists, but merely a group that lost the race to shed their genes and became stuck in the role of function performers.[6]

While Darwin stressed competition and conflict, the Canadian biologist Brian Goodwin* argued that survival is a matter of finding a niche. For Goodwin, organisms that survive are not better than those that have become extinct. Instead evolution is "like a dance" with organisms "simply exploring a space of possibilities."[7]

Summary

While the idea of evolution preceded Darwin, it was his idea of natural selection that made evolution plausible: individuals vary, random selection makes some individuals better suited to their environment than others, individuals better adapted will have disproportionately more offspring. By making evolution the source of different species of animal life rather than God, Darwin finally separated science and religion. By eliminating a Divine Creator, scientific explanations for natural phenomena became paramount.[8] For this innovation, he was, and continues to be, celebrated.

Darwin founded the science of evolutionary biology*, with natural selection as its underlying principle. Another of his contributions to this field was to suggest that species change over time, a process called gradualism*. He also pictured evolution developing in branches, rather than progressing in a linear fashion, implying that all species descend from a single unique origin. While Darwin thought that selection takes place at the level of the individual, we now know it is at the level of the gene. This has recently given rise to a new field of medicine called gene therapy*, which makes it possible to treat genetic disorders, such as cystic fibrosis*, cancer, and certain infectious diseases (including HIV*). Doctors can perform such therapy in utero (in the womb), potentially treating a life-threatening disorder before a child is even born. While gene therapy could spare future generations from particular genetic disorders, it might also affect the development of a fetus in ways we cannot predict, or have long-term side effects that are as yet unknown.[9] Consequently it remains controversial.

1. Office of National Statistics, Mortality Rate in the UK 2010, accessed February 5, 2016, http: // www.ons.gov.uk/ons/rel/mortality-ageing/mortality-in-the-united-kingdom/mortality-in-the-united-kingdom–2010/mortality-in-the-uk-2010.html.

2. Steve Jones, speaking at a lecture marking the bicentenary of Darwin's birth and the 150th anniversary of *On the Origin of Species*, at the University of Cambridge.

3. Jeffrey J. Morris, Richard E. Lenski, and Erik R. Zinser, "The Black Queen Hypothesis: Evolution of Dependencies through Adaptive Gene Loss," *mBio* 3, no. 2 (2012): 1–7.

4. Morris, "The Black Queen Hypothesis," 1–7.

5. Morris, "The Black Queen Hypothesis," 1–7.

6. Morris, "The Black Queen Hypothesis," 1–7.

7. Brian Goodwin, *How the Leopard Changed Its Spots: The Evolution of Complexity* (Princeton New Jersey: Princeton University Press, 2001), 98.

8. Ernst Mayr, "Darwin's Influence on Modern Thought," *Proceedings of the American Philosophical Society* 139, no. 4 (Dec 1995): 317–25.

9. For a review on the issues surrounding the application of gene therapy, see Sonia Y. Hunt, "Controversies in Treatment Approaches: Gene Therapy, IVF, Stem Cells, and Pharmacogenomics," *Nature Education* 1, no. 1 (2008): 222.

GLOSSARY OF TERMS

1. **Abolitionism:** a movement to end slavery in western Europe and the Americas.

2. **Allele:** a variant form of a gene. Some genes come in a variety of different forms.

3. **Altruism:** concern for or devotion to the welfare of others. Behavior that is disadvantageous to the person acting, but beneficial to the recipient, is termed altruistic.

4. **Anatomist:** someone who specializes in animal anatomy.

5. **Anthropology:** the study of human cultural and social life.

6. **Arthropod:** the name given to a group of animals without a backbone, but with a segmented body and jointed limbs. Insects, millipedes, crustaceans, and spiders are all arthropods.

7. **Atheism:** the belief that no god or divine being exists.

8. *Beagle:* the HMS ("Her/His Majesty's Ship") *Beagle* was a Royal Navy vessel on which Charles Darwin spent five years sailing around the world on a scientific voyage, collecting specimens and developing his theory of natural selection.

9. **Behavioral ecology:** a field that uses evolutionary theory and the environment in which an animal lives (including factors such as predators and the weather) to understand behavior.

10. **Biological anthropology:** also referred to as "physical anthropology," this is the biological study of human beings as animals.

11. **Biological determinism:** the suggestion that all (or most) human behavior is dictated by genes, rather than by upbringing or personal choice.

12. **Biological inheritance:** the various characteristics that can be passed from parent to child.

13. **Black Queen hypothesis:** the idea that microorganisms sometimes lose the ability to perform a function that is essential for their survival if other microbes in their immediate environment can perform the function for them. This adaptation encourages microorganisms to live in cooperative communities.

14. **Capitalism:** a social and economic system in which trade and industry are held

in private hands and conducted for private profit.

15. **Comparative anatomy:** the study of the comparison and contrast of the anatomies of different species.

16. **Comparative developmental biology:** a field of inquiry using natural variation and disparity to understand the patterns of growth of life forms at all levels.

17. **Creationism:** the belief that life and the natural world were created by a divinity or divinities.

18. **Cystic fibrosis:** a genetic disorder that compromises the function of organs that are fundamental to life.

19. **Descent with modification:** a term describing how over time and generations, the traits conferring reproductive advantage become more common within a population.

20. **Divergence:** When a species splits to become two or more separate species.

21. **DNA:** an abbreviation for *Deoxyribonucleic acid*, a molecule that carries most of the genetic instructions used in the development, functioning, and reproduction of all living organisms and many viruses.

22. **Ecological niche:** the ecological role of an organism in a community. It applies particularly to food consumption.

23. **Enlightenment:** a period and intellectual movement in Western Europe during the seventeenth and eighteenth centuries. Reason was emphasized over tradition and ideas of divine interference in the affairs of humans. Prominent scholars of the period included the French authors Voltaire and Rousseau, and the German philosopher Immanuel Kant.

24. **Erosion:** the geological process—driven by wind, sun, or water—that breaks down soil or rock structures.

25. **Essentialism:** the view that for any specific entity (such as an animal or a physical object) there is a set of attributes that are necessary to its identity and function. The concept arose from the work of the Greek philosophers Plato and Aristotle.

26. **Eusociality:** a term describing organisms that live in a cooperative group in which usually only one female and several males are reproductively active, the rest being nonbreeding individuals who care for the young, or protect and provide for the group. Examples include termites, ants, and naked mole rats.

27. **Evolution:** the generational changes in heritable traits in organisms.

28. **Evolutionary biology:** the study of species change over time. Subfields in this area include taxonomy, ecology, population genetics, and paleontology.

29. **Evolutionary psychology:** the study of psychology in an evolutionary context.

30. **Evolutionism:** the idea that societies start out as "primitive" and advance toward "civilization".

31. **Extinction:** a biological term for the end of an organism or species.

32. **Gene:** a part of DNA that encodes a functional protein product. It is the molecular unit of heredity.

33. **Gene therapy:** the treatment of certain medical disorders caused by genetic anomalies or deficiencies, by altering or replacing the genes in a patient's cells.

34. **Genesis:** the first book of the Pentateuch (Genesis, Exodus, Leviticus, Numbers, Deuteronomy), part of the Jewish and Christian scriptures. The book describes the creation of the earth and humans, and the expansion of the human race, along with the story of Abraham and his descendants.

35. **Genetic drift:** nonadaptive change that is caused by gene variation due to purely random sampling.

36. **Genetics:** the study of genes and genetic variation in life forms.

37. **Geography:** the study of the planet's land and environments and the ways in which humans interact with their ecological niches and environments.

38. **Geology:** the study of the physical earth, or any celestial body, including fields as diverse as plate tectonics or evolution.

39. **Gradualism:** in evolutionary studies, the concept that species change in intermediate stages over time periods, and generationally.

40. **Great Chain of Being:** a concept first put forward by the ancient Greek philosopher Plato. The main idea is that organisms all belong somewhere along a hierarchy. So everything has its "place," because that is the way God intended it to be.

41. **Great Reform Act:** legislation enacted in England and Wales in 1832 that gave large groups of previously disenfranchised men the right to vote. Similar legislation enacted the same year did the same for Scotland and Ireland.

42. **Group selection:** the process by which an individual acts for the good of the group or species, a theory elaborated upon in 1986 by the English zoologist V. C. Wynne-Edwards that has since been discredited. The concept was first championed by scientists such as the Nobel prize-winning zoologist Konrad Lorenz.

43. **Hamilton's Rule:** explains the conditions required for altruism to evolve, for example when a creature helps its own close relatives to survive at the cost of its own wellbeing. The formula is $r \times B > C$, where r is the degree of relatedness between individuals, B is the benefit to the recipient, and C is the cost for performing that particular act.

44. **HIV:** stands for human immune deficiency virus. This is a retrovirus. The immune system of individuals who are infected has a reduced function. HIV is a cause of AIDS.

45. **Hybridism:** the offspring of two different species, subspecies, or occasionally even genera.

46. **Intellectual priority:** the acknowledgment that a seminal work is the first of its kind and represents a major step forward.

47. **Intelligent design:** a term used to describe the idea that the earth was designed and created by God.

48. **Intersexual selection:** the process that occurs when different sexes are attracted to each other thanks to particular characteristics.

49. **Intrasexual selection:** the process that occurs when members of the same sex compete with each other for opportunities to mate.

50. **Kin selection:** the way certain behaviors are favored over evolutionary time between individuals that are closely related.

51. **Lamarckism:** term in Darwin's time for the transmutation (changing) of species, which went against the official scientific understanding that only fixed states of organisms exist. These days it generally means the incorrect idea that parents develop traits in their lifetime that their offspring will inherit.

52. **Liberalism:** the idea that the government should be responsible for protecting the freedom and equality of individuals in a society.

53. **Life sciences:** the fields of science that deal with living organisms— neuroscience, botany, and virology, for example.

54. **Linguistics:** the study of the structures and nature of language.

55. **Linnean Society:** a society founded in London to promote the study of natural history and taxonomy (the categorization of organisms).

56. **Malthusian catastrophe:** a hypothetical scenario whereby all the members of a society are forced to return to self-sufficient farming in order to provide for their families. This is predicted to happen when population growth has overtaken agricultural production.

57. **Mammal:** the name for a group of animals that have a backbone and where the young are fed by milk produced in the mammary glands. Mammals include dolphins, humans, dogs, and cows, to name a few.

58. **Materialists:** in this sense, those who look for material causes alone for all material phenomena.

59. **Mesopotamian science:** all forms of scholarly inquiry into natural and cultural phenomena, both real and imagined, in the region of ancient Mesopotamia. This includes the invention of writing by the Sumerians, the division of time into the 60–second minute and the 60–minute hour, and perhaps invention of the wheel.

60. **Modern evolutionary biology/modern evolutionary theory/new synthesis:** a reworking of Darwinian evolutionary theory in the 1930s and 1940s to incorporate genetics and other newer theories, such as kin selection and population

demographics.

61. **Monsters:** Darwin called mutations "monsters." Before the work of the pioneering geneticist Gregor Mendel, there was no clear understanding of genetics and inheritance.

62. **Mutation:** a change in the sequence of an organism's genome—that is, its genetic material.

63. **Mutation theory:** the idea that new species are formed not by continuous variations, as Darwin suggested, but by sudden variations, called mutations. First proposed by the biologist Hugo deVries in 1901, it stated that mutations are inherited through successive generations.

64. **Mutualism:** where two organisms of different species cooperate, with each benefitting from the arrangement.

65. **Natural selection:** a mechanism to explain species change. Charles Darwin and the naturalist Alfred Russel Wallace proposed that organisms strive for survival in varying environments; those with traits most favorable to survival live to reproduce and pass on those traits to their offspring, while those species that do not survive long enough to reproduce become extinct.

66. **Naturalist/natural historian:** a scholar of the natural world.

67. **Naturalistic fallacy:** that belief that because something is "natural" it is also "good."

68. **Neo-Darwinism:** a period in time when scientists, namely Wallace and Weismann, came to reject Lamarckian inheritance and promoted natural selection instead as an explanation for how evolution works. The term was coined by George Romanes in 1895.

69. **Nihilism:** the idea that ultimately there is no purpose to life, and so also no purpose to what people do. People who adhere to this philosophical stance also deny the duality of the human body and its soul.

70. **Nonconformists:** also known as Unitarians, this is a seventeenth-century term for people belonging to English Protestant religions that were not the official

Church of England (notably Puritans and Methodists).

71. **Oceanography:** the scientific study of oceans.

72. **Orthogenesis:** the biologist Theodor Eimer's theory that species change as a direct result of some internal force within an individual, which seeks to modify the current type.

73. **Paleontology:** the scientific study of animals and plants that existed in the past and are now preserved in rocks. Scientists working in this field collaborate closely with geologists to identify the period of time when both rock and the fossilized remains of the organism existed.

74. **Pangenesis:** Darwin's theory on inheritance, that has since been proven to be false, whereby information from each cell in the body of an organism travels (in the form of particles) to the reproductive organs; in the reproductive organs all this information is merged and goes to make the sperm and egg.

75. **Paradigm:** an accepted mode of thinking or doing something in a particular way.

76. **Parent species:** the species that gives rise to other species, usually over long time periods. It usually means the most recent common ancestor of two different species.

77. **Phenotype:** the visible characteristics of an individual. The individuals' phenotype is shaped by not only its genes (or its genetic makeup) but also by the environment in which it grows up and lives.

78. **Plinian Society:** A private members' club formed by students at the University of Edinburgh, for the purpose of reading and discussing articles published by scientists on issues to do with natural history.

79. **Positivism:** the principle that knowledge can only advance through the analysis of scientifically verifiable facts.

80. **Primatology:** the study of the order of primates, to which human beings and the other great apes belong.

81. **Reductionism:** different philosophical positions or theories connected to each

other reduced to a "simpler" or more "basic" form.

82. **Relativism:** the idea that no absolute truth exists, and that truth is relative. This is often applied in sociocultural anthropology, which suggests that human behavior and beliefs should be understood in the context of their particular culture.

83. **Scientific priority:** a general term referring to work seminal in a particular area and regarded as a major advancement, such as the discovery of the double helix as the structure of DNA.

84. **Social constructivism:** the idea that behavior is often constructed by the surrounding culture.

85. **Sociobiology:** the field of science investigating the evolutionary origins and function of human social behavior.

86. **Sociocultural anthropology:** the study of human cultural variability, in the present or over particular time periods.

87. **Sexual selection:** the theory that males compete against each other for access to females, and that females (and to a lesser extent males) choose those with whom they wish to mate. As a result, some traits deemed "attractive," such as the peacock's tail, may be passed on to future generations in greater numbers than other traits.

88. **Social Darwinism:** the attempted application of Darwin's mechanism of natural selection to social structures. The implication is that certain races are more "evolved" than others, and that "successful" societies are the fittest, with greater rewards for the strong or most competitive.

89. **Speciation:** a process whereby a new species evolves. This occurs when a single species splits into two or more independent lineages.

90. **Species:** the largest group of organisms in which two individuals are capable of producing fertile offspring, typically using sexual reproduction.

91. **Structuralism:** a theoretical framework used to understand human culture: the patterns and interconnectedness of human interactions are sketched, and the

resulting pattern or structure is used to better understand human culture.

92. **Subspecies:** a group of organisms that can interbreed, and sometimes do so in the wild. In taxonomy, species come after subspecies. This is the first stage of speciation—when new species are formed.

93. **Survival of the fittest:** this term is used to refer to individuals that are best equipped to survive and produce offspring. Thus, in biology, the word "fittest" refers to those individuals that are able to produce the greatest number of offspring during their lifetime.

94. **Taxonomy:** in biology, the systematic classification of living things. Taxonomy and nomenclature (the scientific naming of organisms) was revolutionized by the Swedish naturalist Carl Linnaeus in 1735.

95. **Temperance movement:** a series of mass social and political movements to outlaw or reduce the legality and drinking of alcohol, which reached its peak in terms of political clout in Europe and North America from the mid eighteenth century to the early twentieth century.

96. **Theistic evolution:** the idea that evolution is a process put into action and guided by the hand of God.

97. **Theology:** the systematic study of religious ideas, commonly conducted through religious scripture.

98. **Transmutation (of species):** a phrase coined by the French biologist Jean-Baptiste Lamarck in 1809 to describe the possibility of species change from uniform prototypes.

99. **Uniformitarianism:** the idea that the same forces that change the physical earth today also operated in the past; uniformitarianism was first proposed by the Scottish geologist James Hutton in 1788.

100. **Zoology:** the study of animals.

PEOPLE MENTIONED IN THE TEXT

1. **Sir Charles Bell (1774–1842)** was a Scottish anatomist noted for identifying the function of sensory and motor nerves. He also wrote on religion, philosophy, and theology.

2. **Janet Browne (b. 1950)** is a British historian of science and one of the most acclaimed biographers of Charles Darwin, known for her works *Charles Darwin: Vol. 1, Voyaging* and *Charles Darwin: Vol. 2, The Power of Place*.

3. **Georges-Louis Leclerc, Comte de Buffon (1707–88)** was a French naturalist, mathematician, and director of the Jardin du Roi (later called Jardin de Plantes), a notable botanical garden. He is regarded as the father of natural history during the Enlightenment period. His best-known work, *Histoire Naturelle* (1749–88) is a 36-volume encyclopedia of the animal and mineral kingdoms.

4. **Josep Call (b. 1967)** is a Spanish comparative psychologist. His research focuses on investigating nonhuman primate cognition and comparing it to human intelligence. He is currently the director of the Wolfgang Köhler Primate Research Center at the Max Planck Institute, Leipzig, Germany.

5. **Robert Chambers (1802–71)** was a Scottish scientist and publisher. His pre-Darwinian evolutionary ideas were published anonymously in 1844 in a work called *Vestiges of the Natural History of Creation*.

6. **Carl Erich Correns (1864–1933)** was a German geneticist and botanist. He is best known for discovering the principle of biological heredity, and for rediscovering the work of Gregor Mendel on pea plants and inheritance. He achieved this at the same time as Hugo de Vries, but independently.

7. **Georges Cuvier (1769–1832)** was a French naturalist and zoologist who championed the fixity of species on both religious and scientific grounds. He was best known for *Tableau elementaire de l'histoire naturelle des animaux* (1798).

8. **Emma Darwin (1808–96)**, *née* Wedgwood, became Charles Darwin's wife and helped him raise ten children. She was Darwin's first cousin and they both belonged to the notable Wedgwood family, makers of Wedgwood pottery.

9. **Erasmus Darwin (1731–1802)** was grandfather to Charles Darwin. A prominent scientist and physician, he had himself promoted ideas about the

transmutation of species.

10. **Robert Waring Darwin (1766–1848)** was an English medical doctor, and father of the naturalist Charles Darwin.

11. **Richard Dawkins (b. 1941)** is a science writer, evolutionary biologist, and prominent atheist. He is best known for his 1976 book *The Selfish Gene*, which argues that natural selection takes place at gene level.

12. **Democritus (460 B.C.E.–70 B.C.E.)** was an ancient Greek pre-Socratic philosopher. He is best known for suggesting the atomic theory of the universe. He is considered a materialist, believing everything to be the result of natural laws.

13. **Theodosius Grygorovych Dobzhansky (1900–75)** was a prominent Ukrainian-born geneticist and evolutionary biologist. He later moved to work in the United States. He was a central figure in the field of evolutionary biology and unifying modern evolutionary synthesis.

14. **Gustav Heinrich Theodor Eimer (1843–98)** was a German zoologist. He is credited with popularizing the term orthogenesis to describe evolution guided in a specific direction.

15. **Charles Sutherland Elton (1900–91)** was an English zoologist and ecologist best known for his studies on modern population ecology.

16. **Ronald Fisher (1890–1962)** was a British statistician who incorporated Mendelian genetics into Darwinian theory, resulting in what is now known as the modern evolutionary synthesis. He is one of the key founders of what was later known as population genetics. Less favorably, he was also a prominent eugenicist.

17. **Francis Galton (1822–1911)** was an English anthropologist and statistician, whose achievements included the invention of fingerprint identification. He was also a prominent eugenicist, and the half-cousin of Charles Darwin.

18. **Étienne Geoffroy Saint-Hilaire (1772–1844)** was a French naturalist. He established the principle of unity of composition, suggesting that a single consistent basic plan could be found in all animals.

19. **Brian Carey Goodwin (1931–2009)** was a Canadian mathematician and

biologist. He was one of the founders of theoretical biology—a branch of mathematical biology that uses methods from mathematics and physics to understand processes in biology.

20. **Stephen Jay Gould (1941–2002)** was an American paleontologist, evolutionary biologist, and popular science writer who opposed sociobiological determinism.

21. **Robert Edmond Grant (1793–1874)** was a Scottish doctor and anatomist who taught Darwin at Edinburgh University. Grant later moved to London where he set up the now famous Grant museum of zoology and became the first professor of zoology in Britain. His research on marine invertebrates established that sponges are in fact animals.

22. **William (Bill) D. Hamilton (1936–2000)** was an English evolutionary biologist who theorized that social favoritism shown toward indirect kin (nieces, for example) often appears as altruism when in fact it serves the individual's fitness to favor kin. He is best known for Hamilton's Rule. Hamilton was a forerunner of gene-based evolutionary theorizing from scientists such as Richard Dawkins.

23. **John Stevens Henslow (1796–1861)** was a British botanist, clergyman, and geologist. At the University of Cambridge he introduced a teaching technique that fostered independent discovery, and became a source of inspiration to the young Charles Darwin.

24. **Kim Hill** is an American anthropologist. His work focuses on the evolutionary ecology of human behavior. He is known for his views on human uniqueness, linking humans' success to their sociality.

25. **Joseph Dalton Hooker (1817–1911)** was a well-regarded botanist and explorer, director of the Royal Botanic Gardens, Kew, and close personal friend of Darwin.

26. **James Hutton (1726–97)** was a Scottish geologist best known for his theory of uniformitarianism.

27. **Thomas Henry Huxley (1825–95)** was an anatomist, widely known as "Darwin's bulldog" for his well-publicized defense of Darwin's ideas in *On the Origin of Species*. He wrote *Evidence as to Man's Place in Nature* (1863), where he explicitly stated that humans and apes descended from a common ancestor and

that humans, like other animals, evolved and thus were and are subject to natural selection.

28. **Fleeming Jenkin (1833–1885)** was a British engineer noted for inventing the cable car.

29. **Stephen Jones (b. 1944)** is a Welsh geneticist and evolutionary biologist at University College London.

30. **Stuart Alan Kauffman (b.1939)** is an American theoretical scientist and trained medical doctor. He is best known for arguing that the complexity of biological systems and organisms might result as much from self-organization as from Darwinian natural selection.

31. **Motoo Kimura (1924–94)** was a Japanese biologist who developed the neutral theory of molecular evolution, which suggests that it is drift and not natural selection that is the cause of most species change.

32. **Reverend Charles Kingsley (1819–75)** was an English cleric who vocally spoke and wrote in favor of Darwinian ideas and evolution.

33. **Astrid Kodric-Brown** is an American ecologist and evolutionary biologist who focuses on the behavior of freshwater fish, especially the evolution of mate recognition systems and their role in speciation.

34. **Jean-Baptiste Lamarck (1744–1829)** was a French biologist who suggested the concept of changeable species in his 1809 work *Philosophie Zoologique,* a work that also stated that changes can be inherited generationally.

35. **Richard Lewontin (b. 1929)** is an evolutionary biologist and geneticist best known for his book *Not in Our Genes: Biology, Ideology and Human Nature*, written with Steven Rose and Leon J. Kamin.

36. **Carl Linnaeus (1707–78)** was the Swedish botanist who devised a system for sorting all known organisms, including human beings, into subsets. In his 1735 work *Systema Naturae*, he also invented the still-used taxonomical naming system called binomial nomenclature (which gives living things two names, such as *Homo sapiens*). Linnaeus believed that species (and organisms) are fixed and unchanging.

37. **John Lubbock (1834–1913)** was a mathematician, scientist, banker, politician,

and the 1st Baron of Avebury. Like Darwin, he also lived in the village of Down in Kent and became one of Darwin's closest younger friends.

38. **Charles Lyell (1797–1875)** was a geologist and advocate of uniformitarianism: the idea that the same forces that change the physical earth today also operated in the past. Lyell was a professional and personal supporter of Darwin's ideas.

39. **Thomas Robert Malthus (1766–1834)** was an English demographer and cleric who theorized that human populations are checked by disease and famine. Both Darwin and Wallace developed their respective theories of natural selection after reading Malthus, whose *An Essay on the Principle of Population* was first published in 1798.

40. **Harriet Martineau (1802–76)** was a social theorist and prominent Whig who wrote over 35 notable books on social theory. She was also romantically involved with Darwin's brother Erasmus.

41. **Ernst Mayr (1904–2005)** was a German biologist and Nobel Prize winner who, as well as developing the biological species concept, is considered one of the founders of the modern evolutionary synthesis.

42. **Gregor Mendel (1822–84)** was a Moravian monk whose 1860s writings on genetics and inheritance, based mainly on his work breeding pea plants and widely unknown until 1900, were one of the "missing pieces" of modern evolutionary theory.

43. **Geoffrey Miller (b. 1965)** is an evolutionary psychologist working at the University of New Mexico. His research interests include the study of mate choice in humans.

44. **G. E. Moore (1873–1958)** was an influential British realist philosopher whose systematic approach to ethical problems and remarkably meticulous approach helped found the "analytic" tradition in philosophy. Moore is best known for his defense of ethical non-naturalism and naturalistic philosophy.

45. **Lewis Henry Morgan (1818–81)** was an American anthropologist and a contemporary of Darwin whose theories on the technological "evolution" of various societies became widely circulated for several generations, until challenged by more relativist anthropologists.

46. **Isaac Newton (1642–1726)** was an English physicist who discovered the principles of gravity and laws of motion.

47. **Friedrich Nietzsche (1844–1900)** was a German philosopher noted for philosophical concepts such as the Superman and for his contentious work on the usefulness and purpose of religion.

48. **Noah** is the principal character mentioned in the biblical book of Genesis in the story of Noah's ark, which tells how God destroyed all living things on earth by way of a flood; two of each animal were saved on Noah's ship or ark.

49. **Richard Owen (1804–92)** was an English paleontologist and comparative anatomist probably best known for coining the word *Dinosauria*. He was an outspoken opponent of Darwin's theory of evolution by natural selection—although he agreed that evolution occurred, he believed it was driven by a more complex mechanism than that outlined by Darwin.

50. **Bishop William Paley (1743–1805)** was an English clergyman and philosopher. Using the now famous watchmaker analogy, he argued in his book *Natural Theology or Evidences of the Existence and Attributes of the Deity* that the existence of God is proved by the beauty and complexity of the natural world.

51. **Steven Pinker (b. 1954)** is a Canadian American psychologist well known for his book *The Blank Slate: The Modern Denial of Human Nature*, which takes a biologically determinist stance to human behavior (that is, he believes that our behavior is determined by our biology).

52. **Plato (428–348 B.C.E.)** was a classical Greek philosopher who focused primarily on topics to do with justice, beauty, and equality. He founded the Academy in Athens, the first institution of higher learning in the Western world.

53. **Jean-Jacques Rousseau (1712–78)** was a Swiss writer and philosopher. He wrote *The Social Contract* (1762), which challenged the supremacy of the state—or religious authority—over that of the individual.

54. **Adam Sedgwick (1785–1873)** was a renowned English geologist and Darwin's former geology teacher at Cambridge, who opposed his pupil's theory of evolution.

55. **John Maynard Smith (1920–2004)** was an English evolutionary biologist best

known for his work on game theory in an evolutionary context.

56. **Herbert Spencer (1820–1903)** was an English biologist and anthropologist. A contemporary of Darwin's, he is also known for coining the term "survival of the fittest".

57. **Walter Sutton (1877–1916)** was an American geneticist. Using the work of Gregor Mendel as a starting point, he developed an important theory about chromosomes.

58. **Frederick Temple (1821–1902)** was an English clergyman who became Archbishop of Canterbury. He was known for his interest in the way that religion interacts with science.

59. **Michael Tomasello (b.1950)** is an American-born psychologist, now based in Germany. His research centers on investigating the origin of human intelligence.

60. **Hugo de Vries (1848–1935)** was a Dutch botanist and geneticist. In the 1890s he came up with the concept of genes, after rediscovering the work of Gregor Mendel on heredity in pea plants. He is also notable for introducing the term mutation and developing a theory of evolution based on mutation occurring in genes.

61. **Alfred Russel Wallace (1823–1913)** was a Welsh naturalist. Wallace independently came up with a theory of natural selection that was near-identical to Darwin's as the mechanism for species change. Both Darwin and Wallace are recognized as co-discoverers of the concept of natural selection.

62. **Josiah Wedgwood (1730–95)** was an English pottery designer and manufacturer. He used a scientific approach to pottery making and was known for his exhaustive research into materials.

63. **Friedrich Leopold August Weismann (1834–1914)** was a German evolutionary biologist. He is best known for coming up with the germ plasm theory, which states that (in multicellular organisms) inheritance only takes place by means of the germ cells—the gametes or egg and sperm cells. His idea is vital to modern evolutionary synthesis.

64. **Edward Osborne Wilson (b. 1929)** is an American biologist and environmental activist considered to be the leading thinker and theorist of sociobiology (inquiry into the biological nature of social behavior).

65. **Merlene Zuk (b. 1956)** is an American biologist whose research focuses on sexual selection and mate choice, and the evolution of secondary sex characters.

WORKS CITED

1. Andersson, Malte B. *Sexual Selection: Monographs in Behavior and Ecology*. Princeton: Princeton University Press, 1994.

2. Baden-Powell, Rev. *Philosophy of Creation*. In *Evolution and Dogma* by John Augustine Zahm. Hard Press, 2013.

3. Baguñà, Jaume, and Jordi Garcia-Fernàndez. "*Evo-Devo*: the Long and Winding Road." *The International Journal of Developmental Biology* 47, nos. 7–8 (2003): 705–13.

4. Barton, Nicholas H., Derek E. G. Briggs, Jonathan A. Eisen, David B. Goldstein and Nipam H. Patel. *Evolution*. Cold Spring Harbor, NY: Cold Spring Harbor Laboratory Press, 2007.

5. Bell, Charles. *Essays on the Anatomy and Philosophy of Expression*. Montana: Kessinger Publishing, 2008.

6. Bowler, Peter. *Charles Darwin: The Man and His Influence*. Cambridge: Cambridge University Press, 1996.

7. Bräuer, Juliane, Josep Call, and Michael Tomasello. "Chimpanzees really know what others can see in a competitive situation." *Animal Cognition* 10 (2007): 439–48.

8. Browne, Janet. *Charles Darwin: A Biography. Vol. 1: Voyaging*. Princeton, NJ: Princeton University Press, 1996.

9. *Charles Darwin: A Biography. Vol. 2: The Power of Place*. London: Pimlico, 2003.

10. Buffon, Georges-Louis Leclerc, Comte de. *Les Epoques de la Nature*. In *Histoire Naturelle, générale et particulière, avec la description du Cabinet du Roi*. Paris: imprimerie nationale, 1749–88.

11. Bulmer, Michael. "The theory of natural selection of Alfred Russel Wallace FRS." *Royal Society Journal of the History of Science* 59, no. 2 (2005): 125–36.

12. Cadbury, Deborah. *Terrible Lizard: The First Dinosaur Hunters and the Birth of a New Science*. New York: Henry Holt, 2000.

13. Cañestro, Cristian, Hayato Yokoi, and John H. Postlethwait. "Evolutionary

Developmental Biology and Genomics." *Nature Reviews Genetics* 8, no. 12 (2007): 932–42.

14. Carothers, B. J., and H. T. Reis. "Men and Women Are From Earth: Examining the Latent Structure of Gender." *Journal of Personality and Social Psychology* 104, no. 2 (2013): 385–407.

15. Chambers, Robert. *Vestiges of the Natural History of Creation*. London: John Churchill, 1844.

16. Cook, Peter. *Evolution Versus Intelligent Design: Why All the Fuss? The Arguments for Both Sides*. Australia: New Holland Publishing, 2007.

17. Corning, Peter. *Holistic Darwinism: Synergy, Cybernetics, and the Bioeconomics of Evolution*. Chicago: University of Chicago Press, 2010.

18. Cuvier, G. *Tableau elementaire de l'histoire naturelle des animaux*. Paris: Baudouin, 1798. Accessed February 16, 2016. https: //archive.org/details/ tableaulment00cuvi.

19. Darwin, Charles. *The Descent of Man, and Selection in Relation to Sex*. London: John Murray, 1871.

20. ——. *The Expression of the Emotions in Man and Animals*. London: John Murray, 1872.

21. ——. *On the Origin of Species by Means of Natural Selection, or the Preservation of Favoured Races in the Struggle for Life*. Introduction and notes by Gillian Beer. Oxford: Oxford University Press, 2008.

22. ——. *The Variation of Animals and Plants Under Domestication*. London: John Murray, 1868.

23. Darwin, Charles, and Alfred Russel Wallace. "On the Tendency of Species to form Varieties; and on the Perpetuation of Varieties and Species by Natural Means of Selection." *Journal of the Proceedings of the Linnean Society of London. Zoology* 3 (20 August, 1858): 45–50.

24. Dawkins, Richard. "Science, Delusion and the Appetite for Wonder." *Reports of the National Center for Science Education* 17, no. 1 (1997): 8–14.

25. *The Selfish Gene*. Oxford: Oxford University Press, 1990.

26. Delgado, Cynthia. "Finding Evolution in Medicine." *National Institutes of Health Record* 58, no. 15 (2006).

27. Desmond, Adrian, and James Moore. *Darwin*. London: Michael Joseph, 1991.

28. Durinx, Michel, and Tom J. M. Van Dooren. "Assortative Mate Choice and Dominance Modification: Alternative Ways of Removing Heterozygote Disadvantage." *Evolution* 63, no. 2 (2009): 334–52.

29. *Economist*. "Evolution and Religion: In the Beginning." April 19, 2007. Accessed February 4, 2016. http: //www.economist.com/node/9036706.

30. Eimer, Theodor. *Organic Evolution as the Result of the Inheritance of Acquired Characteristics According to the Laws of Organic Growth*. London: Macmillan, 1890.

31. Eiseley, Loren. *Darwin's Century*. New York: Anchor Books/Doubleday, 1961.

32. Elstein, Daniel. "Species as a Social Construction: Is Species Morally Relevant?" *Journal for Critical Animal Studies* 1, no. 1 (2003): 53–71.

33. Green, Nathan. "Richard Dawkins calls for evolution to be taught to children from age five." *Guardian*, September 1, 2011.

34. Goodwin, Brian. *How the Leopard Changed Its Spots: The Evolution of Complexity*. Princeton, NJ: Princeton University Press, 2001.

35. Gould, S. J. "Impeaching a Self-Appointed Judge." *Scientific American* 267, no. 1 (1992): 118–21.

36. Gould, S. J., and Niles Eldredge. "Punctuated Equilibria: an Alternative to Phyletic Gradualism." In *Models in Paleobiology*, ed. T. J. M. Schopf, 82–115. San Francisco: Freeman Cooper, 1972.

37. Guldberg, Helene. "Restating the Case for Human Uniqueness." *Psychology Today*, November 8, 2010. Accessed February 4, 2016. https: //www. psychologytoday.com/blog/ reclaiming-childhood/201011/restating-the-case-human-uniqueness.

38. Hager, Thomas. *The Life of Linus Pauling*. New York: Simon and Schuster, 1995.

39. Harris, Mark. "Human uniqueness, and are humans the pinnacle of evolution?" *Science and Religion @ Edinburgh*, September 7, 2014. Accessed February 16, 2016. http: //www.blogs.hss.ed.ac.uk/science-and-religion/2014/09/07/human-uniqueness-and-are-humans-the-pinnacle-of-evolution/.

40. Haviland, W. A., and G.W. Crawford. *Human Evolution and Prehistory*. Cambridge, MA: Harvard University Press, 2002.

41. Hawkins, Mike. *Social Darwinism in European and American thought, 1860–1945: Nature as Model and Nature as Threat.* Cambridge: Cambridge University Press, 1998.

42. Hey, Jody. *Genes, Categories and Species*. New York: Oxford University Press 2001.

43. Hollingdale, R. J. *Nietzsche: The Man and His Philosophy*. Cambridge: Cambridge University Press, 1999.

44. Hooke, Samuel Henry. *Middle Eastern Mythology*. Dover Publications, 2013.

45. Hunt, Sonia Y. "Controversies in Treatment Approaches: Gene Therapy, IVF, Stem Cells, and Pharmacogenomics." *Nature Education* 1, no. 1 (2008): 222.

46. Huxley, Thomas Henry. *Huxley Papers*. London: Imperial College of Science and Technology.

47. Hyers, Conrad. *The Meaning of Creation: Genesis and Modern Science*. Louisville: Westminster John Knox, 1984.

48. Joel, Daphna, Zohar Berman, Ido Tavor, Nadav Wexler, Olga Gaber, Yaniv Stein, Nisan Shefi, Jared Pool, Sebastian Urchs, Daniel S. Margulies, Franziskus Liem, Jürgen Hänggi, Lutz Jäncke, and Yaniv Assaf. "Sex beyond the genitalia: the human brain mosaic." *PNAS* 112, no. 50 (2015): 15468–73.

49. Johnson, Phillip E. "The Church of Darwin." *Wall Street Journal,* August 16, 1999.

50. Jones, Steve. *Darwin's Island: The Galapagos in the Garden of England*. London: Little Brown, 2009.

51. Kitcher, Philip. *Living with Darwin: Evolution, Design, and the Future of Faith.*

New York; Oxford: Oxford University Press, 2007.

52. Krebs, J. R., and A. Kacelnik. *Behavioural Ecology: An Evolutionary Approach*. Oxford: Blackwell Scientific, 1991.

53. Johnson, Phillip E. *Darwin on Trial*. Downers Grove, IL: InterVarsity Press, 1991.

54. Lamarck, J. B. *Zoological Philosophy: An Exposition with Regard to the Natural History of Animals*. Translated by Hugh Elliot. Chicago: University of Chicago Press, 1984.

55. Larson, Edward J. E. *Evolution: The Remarkable History of a Scientific Theory*. New York: Modern Library, 2004.

56. Leff, David. "About Charles Darwin." Accessed February 4, 2016. www. aboutDarwin.com.

57. Levine, George. *Darwin the Writer*. Oxford: Oxford University Press, 2011.

58. Lewontin, R. C. "Biological Determinism." The Tanner Lectures on Human Values, University of Utah, March 31 and April 1, 1982. Accessed February 5, 2016. http: // tannerlectures.utah.edu/_documents/a-to-z/l/lewontin83.pdf.

59. Lyell, Charles. *The Principles of Geology: Being an Attempt to Explain the Former Changes of the Earth's Surface, by Reference to Causes Now in Operation*. 3 Volumes. London: John Murray, 1830–3.

60. Magner, Lois N. *A History of the Life Sciences*. New York; Basel: Marcel Dekker, 1994.

61. Mallet, J. A. "Species definition for the modern synthesis." *Trends in Ecology and Evolution* 10 (1995): 294–9.

62. Mayr, Ernst. "Darwin's Influence on Modern Thought." *Proceedings of the American Philosophical Society* 139, no. 4 (Dec 1995): 317–25.

63. Mendel, Gregor. "Versuche über Pflanzenhybriden." In *Verhandlungen des naturforschenden Vereins in Brünn*, 1866.

64. Miller, David. "Natural Selection and its Scientific Status." *Popper Selections*, ed. David Miller. Princeton, NJ: Princeton University Press, 1985.

65. Moore, G. E. *Principia Ethica*. Cambridge: Cambridge University Press, 1993.

66. Morgan, Thomas Hunt, A. H. Sturtevant, H. J. Muller, and C. B. Bridges. *The Mechanism of Mendelian Heredity*. New York: Henry Holt and Company, 1915.

67. Morris, Jeffrey J., Richard E. Lenski, and Erik R. Zinser. "The Black Queen Hypothesis: Evolution of Dependencies Through Adaptive Gene Loss." *MBio* 3, no. 2 (2012): 1–7.

68. Nowak, Martin A. "Five rules for the evolution of cooperation." *Science* 314 (2006): 1560–3.

69. Ollerton, J. "Speciation: Flowering time and the Wallace Effect." *Heredity* 95, no. 3 (2005): 181–2.

70. Orr, H. Allen. "Testing Natural Selection with Genetics." *Scientific American* 300, no. 1 (2009): 44.

71. Paley, William. *Natural Theology: or, Evidences of the Existence and Attributes of the Deity*. London: J. Faulder, 1809.

72. Pew Forum on Religion and Public Life. "Religious Groups: Opinions of Evolution." February 4, 2009. Accessed February 5, 2016. http: //www. pewforum.org/2009/02/04/religious-differences-on-the-question-of-evolution/.

73. Popper, Karl. "Natural Selection and its Scientific Status." *Popper Selections*, ed. David Miller, 241–43. Princeton, NJ: Princeton University Press, 1985.

74. Reed, Edward S. "The Lawfulness of Natural Selection." *American Naturalist* 118, no. 1 (1981): 61–71.

75. Ridley, Mark. *How to Read Darwin*. London: Granta Books, 2006.

76. Ridley, Matt. "Darwin's Legacy: Modern Darwins." *National Geographic*, February 2009.

77. Robinson, Bruce A. "U.S. public opinion polls about evolution & creation." Accessed February 16, 2016. http: //www.religioustolerance.org/ev_publi2007. htm.

78. Rutter, Michael. *Genes and Behavior: Nature-Nurture Interplay Explained*. Oxford: Blackwell, 2006.

6

79. Sachs, Joel L. "Cooperation Within and Among species." *Journal of Evolutionary Biology* 19, no. 5 (2006): 1415–8.

80. Secord, James A. *Victorian Sensation: The Extraordinary Publication, Reception, and Secret Authorship of Vestiges of the Natural History of Creation.* Chicago: University of Chicago Press, 2000.

81. Sedgwick, Adam. "Review of Vestiges." *Edinburgh Review* 82 (July 1845): 1–85.

82. Sewell, Dennis. *The Political Gene: How Darwin's Ideas Changed Politics.* London: Picador, 2009.

83. Slevin, Peter. "Battle on Teaching Evolution Sharpens." *Washington Post*, March 14, 2005.

84. Smith, Charles. H. "Wallace's Unfinished Business: The 'Other Man' in Evolutionary Theory." *Complexity* 10, no 2 (2004): 25–32.

85. Stott, Rebecca. *Darwin's Ghosts: In Search of the First Evolutionists.* London: Bloomsbury, 2012.

86. Strobel, Lee. *The Case for a Creator: A Journalist Investigates Scientific Evidence That Points Toward God.* Grand Rapids, MI: Zondervan, 2004.

87. Tanner, Julia. "The Naturalistic Fallacy." *Richmond Journal of Philosophy* 13 (2006): 1–6.

88. Tomasello, Michael, Malinda Carpenter, Josep Call, Tanya Behne, and Henrike Moll. "Understanding and sharing intentions: the origins of cultural cognition." *Behavioral and Brain Sciences* 28, no. 5 (2005): 675–735.

89. Wallace, Alfred Russel. *Darwinism: An Exposition of the Theory of Natural Selection, with Some of Its Applications.* London: Macmillan & Co, 1889.

90. "Letter to George Silk." Natural History Museum. Wallace Letters Online: Letter WCP373.373.

91. *My Life: A Record of Events and Opinions.* London: Chapman and Hall, 1905.

92. Weikart, R. *From Darwin to Hitler: Evolutionary Ethics, Eugenics and Racism in Germany.* London: Palgrave Macmillan; 2004.

93. Weismann, Friedrich Leopold August. *Die Entstehung der Sexualzellen bei den Hydromedusen: Zugleich ein Beitrag zur Kenntniss des Baues und der Lebenserscheinungen dieser Gruppe*. Jena: Fischer, 1883.

94. Wildman, Derek. E., Monica Uddin, Guozhen Liu, Lawrence I. Grossman, and Morris Goodman. "Implications of natural selection in shaping 99.4% nonsynonymous DNA identity between humans and chimpanzees: Enlarging genus Homo." *Proceedings of the National Academy of Sciences of the United States of America (PNAS)* 100, no. 12 (2003): 7181–8.

95. Wilkins, John. *Species: A History of the Idea*. Oakland, CA: University of California Press, 2011.

96. Williams, George C. "Pleiotropy, natural selection, and the evolution of senescence," *Evolution* 11, no. 4 (1957): 398–411.

97. Wilson, E. O. "Human Decency Is Animal." *New York Times Magazine*, October 12, 1975.

98. Witham, Larry. "Many Scientists See God's Hand in Evolution." *Reports of the National Center for Science Education* 17, no. 6 (November–December 1997): 33.

99. Wyhe, J. van. "Mind the Gap: Did Darwin Avoid Publishing His Theory For Many Years?" *Royal Society Journal of the History of Science* 61, no. 2 (2007): 177–205.

100. Wynne-Edwards, V. C. *Evolution Through Group Selection*. Oxford: Blackwell Scientific, 2006.

原书作者简介

查尔斯·达尔文 1809 年生于英国什鲁斯伯里镇，家境优渥、社会关系良好，家族成员内有多位医生和科学家。达尔文于 1831 年毕业于剑桥大学基督学院，数月后谋到一个随船博物学家的职位，登上了皇家海军贝格尔号，开始了考察南美洲的航行之旅。

这次航行被证明是达尔文生活的转折点。他利用沿途收集到的信息，提出了一个用来解释进化过程的理论，他称其为"自然选择"。达尔文于 1859 年发表了《物种起源》，并陆续撰写了另外 25 部著作。查尔斯·达尔文于 1882 年去世，葬于伦敦威斯敏斯特大教堂。

本书作者简介

凯瑟琳·布莱森，现于伦敦大学学院攻读进化人类学博士学位，研究重点是人类和其他类人猿中偏见和歧视的认知及适应性根源。

娜迪达·约瑟芬·姆辛达伊，现于伦敦大学学院攻读进化人类学博士学位，研究内容是引进的黑猩猩种群中的行为构建问题。

世界名著中的批判性思维

《世界思想宝库钥匙丛书》致力于深入浅出地阐释全世界著名思想家的观点，不论是谁、在何处都能了解到，从而推进批判性思维发展。

《世界思想宝库钥匙丛书》与世界顶尖大学的一流学者合作，为一系列学科中最有影响的著作推出新的分析文本，介绍其观点和影响。在这一不断扩展的系列中，每种选入的著作都代表了历经时间考验的思想典范。通过为这些著作提供必要背景、揭示原作者的学术渊源以及说明这些著作所产生的影响，本系列图书希望让读者以新视角看待这些划时代的经典之作。读者应学会思考、运用并挑战这些著作中的观点，而不是简单接受它们。

ABOUT THE AUTHOR OF THE ORIGINAL WORK

Charles Darwin was born in 1809 in Shrewsbury, England. He was a member of a wealthy and well-connected family, which included many physicians and scientists. Darwin graduated from Christ's College, Cambridge, in 1831, and a few months later took a position as ship's naturalist on board HMS *Beagle*, a Royal Navy ship then embarking on a surveying voyage to South America.

The voyage proved to be a turning point in Darwin's life. Using the information he collected along the way, he was able to formulate a theory to explain how evolution works. He called this "natural selection." Darwin published *On the Origin of Species* in 1859 and went on to write another 25 books. Charles Darwin died in 1882 and is buried in Westminster Abbey, London.

ABOUT THE AUTHORS OF THE ANALYSIS

Kathleen Bryson is a PhD candidate in evolutionary anthropology at University College, London, where the focus of her research is on the cognitive and adaptative roots of prejudice and discrimination in humans and other apes.

Nadezda Josephine Msindai is a PhD candidate in evolutionary anthropology at University College, London, where she is examining construction behaviour in an introduced population of chimpanzees.

ABOUT MACAT

GREAT WORKS FOR CRITICAL THINKING

Macat is focused on making the ideas of the world's great thinkers accessible and comprehensible to everybody, everywhere, in ways that promote the development of enhanced critical thinking skills.

It works with leading academics from the world's top universities to produce new analyses that focus on the ideas and the impact of the most influential works ever written across a wide variety of academic disciplines. Each of the works that sit at the heart of its growing library is an enduring example of great thinking. But by setting them in context — and looking at the influences that shaped their authors, as well as the responses they provoked — Macat encourages readers to look at these classics and game-changers with fresh eyes. Readers learn to think, engage and challenge their ideas, rather than simply accepting them.

批判性思维与《物种起源》

首要的批判性思维技巧：创造性思维

次要的批判性思考技巧：解决问题

在《物种起源》一书中，查尔斯·达尔文将各种批判性思维技巧广泛而有力地结合起来，对生物变化进行了广泛的解释。

《物种起源》同时也是一本罕见的书，它着力解决不同物种间存在的巨大多样性的问题，并尝试引用大量证据解决这一问题。不过，也许正是达尔文高超的创造才能造就了这部杰作，正是因为这一点，他才得以在迥然不同领域中的众多不同的证据之间建立崭新的联系。

然而，在这本书里撰写的几十年过程中，达尔文的批判性思维能力又是必要的。从整体上看，达尔文对自己提出的问题进行了仔细的研究，对各种解释进行了反复推敲，并且用严密的论证来证明结论的合理性。1859 年在该书出版之际，对于达尔文和其他研究人员所观察到的变化，存在着各种不同的解释。与同时代其他人不同的是，达尔文运用批判性思维，找到了一种全新的方法，将解释与证据有机结合。迄今为止，这种方法优雅、完整，且具有预见性。

CRITICAL THINKING AND *ON THE ORIGIN OF SPECIES*

- Primary critical thinking skill: CREATIVE THINKING
- Secondary critical thinking skill: PROBLEM-SOLVING

Charles Darwin called on a broad and unusually powerful combination of critical thinking skills to create his wide-ranging explanation for biological variety, *On the Origin of Species*.

It's one of those rare books that takes a huge problem—the enormous diversity of different species—and seeks to use a vast range of evidence to solve it. But it was perhaps Darwin's towering creative prowess that made the most telling contribution to this masterpiece, for it was this that enabled him to make the necessary fresh connections between so much disparate evidence from such a diversity of fields.

All of Darwin's critical thinking skills were required, however, in the course of the decades of work that went into this volume. Taken as a whole, Darwin's solution to the problem that he set himself is carefully researched, considers multiple explanations, and justifies its conclusions with well-organised reasoning. At the time of the publication, in 1859, there were various explanations for the changes that Darwin—and others—observed; what separated Darwin from so many of his contemporaries is that he deployed critical thinking to arrive at a significantly new way of fitting explanation to evidence; one that remains elegant, complete and predictive to this day.

《世界思想宝库钥匙丛书》简介

《世界思想宝库钥匙丛书》致力于为一系列在各领域产生重大影响的人文社科类经典著作提供独特的学术探讨。每一本读物都不仅仅是原经典著作的内容摘要，而是介绍并深入研究原经典著作的学术渊源、主要观点和历史影响。这一丛书的目的是提供一套学习资料，以促进读者掌握批判性思维，从而更全面、深刻地去理解重要思想。

每一本读物分为3个部分：学术渊源、学术思想和学术影响，每个部分下有4个小节。这些章节旨在从各个方面研究原经典著作及其反响。

由于独特的体例，每一本读物不但易于阅读，而且另有一项优点：所有读物的编排体例相同，读者在进行某个知识层面的调查或研究时可交叉参阅多本该丛书中的相关读物，从而开启跨领域研究的路径。

为了方便阅读，每本读物最后还列出了术语表和人名表（在书中则以星号＊标记），此外还有参考文献。

《世界思想宝库钥匙丛书》与剑桥大学合作，理清了批判性思维的要点，即如何通过6种技能来进行有效思考。其中3种技能让我们能够理解问题，另3种技能让我们有能力解决问题。这6种技能合称为"批判性思维PACIER模式"，它们是：

分析：了解如何建立一个观点；

评估：研究一个观点的优点和缺点；

阐释：对意义所产生的问题加以理解；

创造性思维：提出新的见解，发现新的联系；

解决问题：提出切实有效的解决办法；

理性化思维：创建有说服力的观点。

THE MACAT LIBRARY

The Macat Library is a series of unique academic explorations of seminal works in the humanities and social sciences — books and papers that have had a significant and widely recognised impact on their disciplines. It has been created to serve as much more than just a summary of what lies between the covers of a great book. It illuminates and explores the influences on, ideas of, and impact of that book. Our goal is to offer a learning resource that encourages critical thinking and fosters a better, deeper understanding of important ideas.

Each publication is divided into three Sections: Influences, Ideas, and Impact. Each Section has four Modules. These explore every important facet of the work, and the responses to it.

This Section-Module structure makes a Macat Library book easy to use, but it has another important feature. Because each Macat book is written to the same format, it is possible (and encouraged!) to cross-reference multiple Macat books along the same lines of inquiry or research. This allows the reader to open up interesting interdisciplinary pathways.

To further aid your reading, lists of glossary terms and people mentioned are included at the end of this book (these are indicated by an asterisk [*] throughout) — as well as a list of works cited.

Macat has worked with the University of Cambridge to identify the elements of critical thinking and understand the ways in which six different skills combine to enable effective thinking.

Three allow us to fully understand a problem; three more give us the tools to solve it. Together, these six skills make up the PACIER model of critical thinking. They are:

ANALYSIS — understanding how an argument is built
EVALUATION — exploring the strengths and weaknesses of an argument
INTERPRETATION — understanding issues of meaning
CREATIVE THINKING — coming up with new ideas and fresh connections
PROBLEM-SOLVING — producing strong solutions
REASONING — creating strong arguments

"《世界思想宝库钥匙丛书》提供了独一无二的跨学科学习和研究工具。它介绍那些革新了各自学科研究的经典著作，还邀请全世界一流专家和教育机构进行严谨的分析，为每位读者打开世界顶级教育的大门。"

—— 安德烈亚斯·施莱歇尔，
经济合作与发展组织教育与技能司司长

"《世界思想宝库钥匙丛书》直面大学教育的巨大挑战……他们组建了一支精干而活跃的学者队伍，来推出在研究广度上颇具新意的教学材料。"

—— 布罗尔斯教授、勋爵，剑桥大学前校长

"《世界思想宝库钥匙丛书》的愿景令人赞叹。它通过分析和阐释那些曾深刻影响人类思想以及社会、经济发展的经典文本，提供了新的学习方法。它推动批判性思维，这对于任何社会和经济体来说都是至关重要的。这就是未来的学习方法。"

—— 查尔斯·克拉克阁下，英国前教育大臣

"对于那些影响了各自领域的著作，《世界思想宝库钥匙丛书》能让人们立即了解到围绕那些著作展开的评论性言论，这让该系列图书成为在这些领域从事研究的师生们不可或缺的资源。"

—— 威廉·特朗佐教授，加利福尼亚大学圣地亚哥分校

"Macat offers an amazing first-of-its-kind tool for interdisciplinary learning and research. Its focus on works that transformed their disciplines and its rigorous approach, drawing on the world's leading experts and educational institutions, opens up a world-class education to anyone."

—— Andreas Schleicher, Director for Education and Skills,
Organisation for Economic Co-operation and Development

"Macat is taking on some of the major challenges in university education... They have drawn together a strong team of active academics who are producing teaching materials that are novel in the breadth of their approach."

—— Prof Lord Broers, former Vice-Chancellor of the University of Cambridge

"The Macat vision is exceptionally exciting. It focuses upon new modes of learning which analyse and explain seminal texts which have profoundly influenced world thinking and so social and economic development. It promotes the kind of critical thinking which is essential for any society and economy. This is the learning of the future."

—— Rt Hon Charles Clarke, former UK Secretary of State for Education

"The Macat analyses provide immediate access to the critical conversation surrounding the books that have shaped their respective discipline, which will make them an invaluable resource to all of those, students and teachers, working in the field."

—— Prof William Tronzo, University of California at San Diego

TITLE	中文书名	类别
An Analysis of Arjun Appadurai's *Modernity at Large: Cultural Dimensions of Globalization*	解析阿尔君·阿帕杜莱《消失的现代性：全球化的文化维度》	人类学
An Analysis of Claude Lévi-Strauss's *Structural Anthropology*	解析克劳德·列维-斯特劳斯《结构人类学》	人类学
An Analysis of Marcel Mauss's *The Gift*	解析马塞尔·莫斯《礼物》	人类学
An Analysis of Jared M. Diamond's *Guns, Germs, and Steel: The Fate of Human Societies*	解析贾雷德·M.戴蒙德《枪炮、病菌与钢铁：人类社会的命运》	人类学
An Analysis of Clifford Geertz's *The Interpretation of Cultures*	解析克利福德·格尔茨《文化的解释》	人类学
An Analysis of Philippe Ariès's *Centuries of Childhood: A Social History of Family Life*	解析菲力浦·阿利埃斯《儿童的世纪：旧制度下的儿童和家庭生活》	人类学
An Analysis of W. Chan Kim & Renée Mauborgne's *Blue Ocean Strategy*	解析金伟灿/勒妮·莫博涅《蓝海战略》	商业
An Analysis of John P. Kotter's *Leading Change*	解析约翰·P.科特《领导变革》	商业
An Analysis of Michael E. Porter's *Competitive Strategy: Techniques for Analyzing Industries and Competitors*	解析迈克尔·E.波特《竞争战略：分析产业和竞争对手的技术》	商业
An Analysis of Jean Lave & Etienne Wenger's *Situated Learning: Legitimate Peripheral Participation*	解析琼·莱夫/艾蒂纳·温格《情境学习：合法的边缘性参与》	商业
An Analysis of Douglas McGregor's *The Human Side of Enterprise*	解析道格拉斯·麦格雷戈《企业的人性面》	商业
An Analysis of Milton Friedman's *Capitalism and Freedom*	解析米尔顿·弗里德曼《资本主义与自由》	商业
An Analysis of Ludwig von Mises's *The Theory of Money and Credit*	解析路德维希·冯·米塞斯《货币和信用理论》	经济学
An Analysis of Adam Smith's *The Wealth of Nations*	解析亚当·斯密《国富论》	经济学
An Analysis of Thomas Piketty's *Capital in the Twenty-First Century*	解析托马斯·皮凯蒂《21世纪资本论》	经济学
An Analysis of Nassim Nicholas Taleb's *The Black Swan: The Impact of the Highly Improbable*	解析纳西姆·尼古拉斯·塔勒布《黑天鹅：如何应对不可预知的未来》	经济学
An Analysis of Ha-Joon Chang's *Kicking Away the Ladder*	解析张夏准《富国陷阱：发达国家为何踢开梯子》	经济学
An Analysis of Thomas Robert Malthus's *An Essay on the Principle of Population*	解析托马斯·罗伯特·马尔萨斯《人口论》	经济学

An Analysis of John Maynard Keynes's *The General Theory of Employment, Interest and Money*	解析约翰·梅纳德·凯恩斯《就业、利息和货币通论》	经济学
An Analysis of Milton Friedman's *The Role of Monetary Policy*	解析米尔顿·弗里德曼《货币政策的作用》	经济学
An Analysis of Burton G. Malkiel's *A Random Walk Down Wall Street*	解析伯顿·G.马尔基尔《漫步华尔街》	经济学
An Analysis of Friedrich A. Hayek's *The Road to Serfdom*	解析弗里德里希·A.哈耶克《通往奴役之路》	经济学
An Analysis of Charles P. Kindleberger's *Manias, Panics, and Crashes: A History of Financial Crises*	解析查尔斯·P.金德尔伯格《疯狂、惊恐和崩溃：金融危机史》	经济学
An Analysis of Amartya Sen's *Development as Freedom*	解析阿马蒂亚·森《以自由看待发展》	经济学
An Analysis of Rachel Carson's *Silent Spring*	解析蕾切尔·卡森《寂静的春天》	地理学
An Analysis of Charles Darwin's *On the Origin of Species: by Means of Natural Selection, or The Preservation of Favoured Races in the Struggle for Life*	解析查尔斯·达尔文《物种起源》	地理学
An Analysis of World Commission on Environment and Development's *The Brundtland Report: Our Common Future*	解析世界环境与发展委员会《布伦特兰报告：我们共同的未来》	地理学
An Analysis of James E. Lovelock's *Gaia: A New Look at Life on Earth*	解析詹姆斯·E.拉伍洛克《盖娅：地球生命的新视野》	地理学
An Analysis of Paul Kennedy's *The Rise and Fall of the Great Powers: Economic Change and Military Conflict from 1500–2000*	解析保罗·肯尼迪《大国的兴衰：1500—2000年的经济变革与军事冲突》	历史
An Analysis of Janet L. Abu-Lughod's *Before European Hegemony: The World System A. D. 1250–1350*	解析珍妮特·L.阿布-卢格霍德《欧洲霸权之前：1250—1350年的世界体系》	历史
An Analysis of Alfred W. Crosby's *The Columbian Exchange: Biological and Cultural Consequences of 1492*	解析艾尔弗雷德·W.克罗斯比《哥伦布大交换：1492年以后的生物影响和文化冲击》	历史
An Analysis of Tony Judt's *Postwar: A History of Europe since 1945*	解析托尼·朱特《战后欧洲史》	历史
An Analysis of Richard J. Evans's *In Defence of History*	解析理查德·J.艾文斯《捍卫历史》	历史
An Analysis of Eric Hobsbawm's *The Age of Revolution: Europe 1789–1848*	解析艾瑞克·霍布斯鲍姆《革命的年代：欧洲1789—1848年》	历史

An Analysis of Roland Barthes's *Mythologies*	解析罗兰·巴特《神话学》	文学与批判理论
An Analysis of Simone de Beauvoir's *The Second Sex*	解析西蒙娜·德·波伏娃《第二性》	文学与批判理论
An Analysis of Edward W. Said's *Orientalism*	解析爱德华·W. 萨义德《东方主义》	文学与批判理论
An Analysis of Virginia Woolf's *A Room of One's Own*	解析弗吉尼亚·伍尔芙《一间自己的房间》	文学与批判理论
An Analysis of Judith Butler's *Gender Trouble*	解析朱迪斯·巴特勒《性别麻烦》	文学与批判理论
An Analysis of Ferdinand de Saussure's *Course in General Linguistics*	解析费尔迪南·德·索绪尔《普通语言学教程》	文学与批判理论
An Analysis of Susan Sontag's *On Photography*	解析苏珊·桑塔格《论摄影》	文学与批判理论
An Analysis of Walter Benjamin's *The Work of Art in the Age of Mechanical Reproduction*	解析瓦尔特·本雅明《机械复制时代的艺术作品》	文学与批判理论
An Analysis of W. E. B. Du Bois's *The Souls of Black Folk*	解析 W.E.B. 杜波依斯《黑人的灵魂》	文学与批判理论
An Analysis of Plato's *The Republic*	解析柏拉图《理想国》	哲学
An Analysis of Plato's *Symposium*	解析柏拉图《会饮篇》	哲学
An Analysis of Aristotle's *Metaphysics*	解析亚里士多德《形而上学》	哲学
An Analysis of Aristotle's *Nicomachean Ethics*	解析亚里士多德《尼各马可伦理学》	哲学
An Analysis of Immanuel Kant's *Critique of Pure Reason*	解析伊曼努尔·康德《纯粹理性批判》	哲学
An Analysis of Ludwig Wittgenstein's *Philosophical Investigations*	解析路德维希·维特根斯坦《哲学研究》	哲学
An Analysis of G. W. F. Hegel's *Phenomenology of Spirit*	解析 G. W. F. 黑格尔《精神现象学》	哲学
An Analysis of Baruch Spinoza's *Ethics*	解析巴鲁赫·斯宾诺莎《伦理学》	哲学
An Analysis of Hannah Arendt's *The Human Condition*	解析汉娜·阿伦特《人的境况》	哲学
An Analysis of G. E. M. Anscombe's *Modern Moral Philosophy*	解析 G. E. M. 安斯康姆《现代道德哲学》	哲学
An Analysis of David Hume's *An Enquiry Concerning Human Understanding*	解析大卫·休谟《人类理解研究》	哲学

An Analysis of Søren Kierkegaard's *Fear and Trembling*	解析索伦·克尔凯郭尔《恐惧与战栗》	哲学
An Analysis of René Descartes's *Meditations on First Philosophy*	解析勒内·笛卡尔《第一哲学沉思录》	哲学
An Analysis of Friedrich Nietzsche's *On the Genealogy of Morality*	解析弗里德里希·尼采《论道德的谱系》	哲学
An Analysis of Gilbert Ryle's *The Concept of Mind*	解析吉尔伯特·赖尔《心的概念》	哲学
An Analysis of Thomas Kuhn's *The Structure of Scientific Revolutions*	解析托马斯·库恩《科学革命的结构》	哲学
An Analysis of John Stuart Mill's *Utilitarianism*	解析约翰·斯图亚特·穆勒《功利主义》	哲学
An Analysis of Aristotle's *Politics*	解析亚里士多德《政治学》	政治学
An Analysis of Niccolò Machiavelli's *The Prince*	解析尼科洛·马基雅维利《君主论》	政治学
An Analysis of Karl Marx's *Capital*	解析卡尔·马克思《资本论》	政治学
An Analysis of Benedict Anderson's *Imagined Communities*	解析本尼迪克特·安德森《想象的共同体》	政治学
An Analysis of Samuel P. Huntington's *The Clash of Civilizations and the Remaking of World Order*	解析塞缪尔·P.亨廷顿《文明的冲突与世界秩序的重建》	政治学
An Analysis of Alexis de Tocqueville's *Democracy in America*	解析阿列克西·德·托克维尔《论美国的民主》	政治学
An Analysis of John A. Hobson's *Imperialism: A Study*	解析约翰·A.霍布森《帝国主义》	政治学
An Analysis of Thomas Paine's *Common Sense*	解析托马斯·潘恩《常识》	政治学
An Analysis of John Rawls's *A Theory of Justice*	解析约翰·罗尔斯《正义论》	政治学
An Analysis of Francis Fukuyama's *The End of History and the Last Man*	解析弗朗西斯·福山《历史的终结与最后的人》	政治学
An Analysis of John Locke's *Two Treatises of Government*	解析约翰·洛克《政府论》	政治学
An Analysis of Sun Tzu's *The Art of War*	解析孙武《孙子兵法》	政治学
An Analysis of Henry Kissinger's *World Order: Reflections on the Character of Nations and the Course of History*	解析亨利·基辛格《世界秩序》	政治学
An Analysis of Jean-Jacques Rousseau's *The Social Contract*	解析让-雅克·卢梭《社会契约论》	政治学

An Analysis of Odd Arne Westad's *The Global Cold War: Third World Interventions and the Making of Our Times*	解析文安立《全球冷战：美苏对第三世界的干涉与当代世界的形成》	政治学
An Analysis of Sigmund Freud's *The Interpretation of Dreams*	解析西格蒙德·弗洛伊德《梦的解析》	心理学
An Analysis of William James' *The Principles of Psychology*	解析威廉·詹姆斯《心理学原理》	心理学
An Analysis of Philip Zimbardo's *The Lucifer Effect*	解析菲利普·津巴多《路西法效应》	心理学
An Analysis of Leon Festinger's *A Theory of Cognitive Dissonance*	解析利昂·费斯汀格《认知失调论》	心理学
An Analysis of Richard H. Thaler & Cass R. Sunstein's *Nudge: Improving Decisions about Health, Wealth, and Happiness*	解析理查德·H. 泰勒/卡斯·R. 桑斯坦《助推：如何做出有关健康、财富和幸福的更优决策》	心理学
An Analysis of Gordon Allport's *The Nature of Prejudice*	解析高尔登·奥尔波特《偏见的本质》	心理学
An Analysis of Steven Pinker's *The Better Angels of Our Nature: Why Violence Has Declined*	解析斯蒂芬·平克《人性中的善良天使：暴力为什么会减少》	心理学
An Analysis of Stanley Milgram's *Obedience to Authority*	解析斯坦利·米尔格拉姆《对权威的服从》	心理学
An Analysis of Betty Friedan's *The Feminine Mystique*	解析贝蒂·弗里丹《女性的奥秘》	心理学
An Analysis of David Riesman's *The Lonely Crowd: A Study of the Changing American Character*	解析大卫·理斯曼《孤独的人群：美国人社会性格演变之研究》	社会学
An Analysis of Franz Boas's *Race, Language and Culture*	解析弗朗兹·博厄斯《种族、语言与文化》	社会学
An Analysis of Pierre Bourdieu's *Outline of a Theory of Practice*	解析皮埃尔·布尔迪厄《实践理论大纲》	社会学
An Analysis of Max Weber's *The Protestant Ethic and the Spirit of Capitalism*	解析马克斯·韦伯《新教伦理与资本主义精神》	社会学
An Analysis of Jane Jacobs's *The Death and Life of Great American Cities*	解析简·雅各布斯《美国大城市的死与生》	社会学
An Analysis of C. Wright Mills's *The Sociological Imagination*	解析 C. 赖特·米尔斯《社会学的想象力》	社会学
An Analysis of Robert E. Lucas Jr.'s *Why Doesn't Capital Flow from Rich to Poor Countries?*	解析小罗伯特·E. 卢卡斯《为何资本不从富国流向穷国？》	社会学

An Analysis of Émile Durkheim's *On Suicide*	解析埃米尔·迪尔凯姆《自杀论》	社会学
An Analysis of Eric Hoffer's *The True Believer: Thoughts on the Nature of Mass Movements*	解析埃里克·霍弗《狂热分子：群众运动圣经》	社会学
An Analysis of Jared M. Diamond's *Collapse: How Societies Choose to Fail or Survive*	解析贾雷德·M.戴蒙德《大崩溃：社会如何选择兴亡》	社会学
An Analysis of Michel Foucault's *The History of Sexuality Vol. 1: The Will to Knowledge*	解析米歇尔·福柯《性史（第一卷）：求知意志》	社会学
An Analysis of Michel Foucault's *Discipline and Punish*	解析米歇尔·福柯《规训与惩罚》	社会学
An Analysis of Richard Dawkins's *The Selfish Gene*	解析理查德·道金斯《自私的基因》	社会学
An Analysis of Antonio Gramsci's *Prison Notebooks*	解析安东尼奥·葛兰西《狱中札记》	社会学
An Analysis of Augustine's *Confessions*	解析奥古斯丁《忏悔录》	神学
An Analysis of C. S. Lewis's *The Abolition of Man*	解析 C. S. 路易斯《人之废》	神学

图书在版编目（CIP）数据

解析查尔斯·达尔文《物种起源》：汉、英 / 凯瑟琳·布莱森（Kathleen Bryson），娜迪达·约瑟芬·姆辛达伊（Nadezda Josephine Msindai）著；贾顺厚译. —上海：上海外语教育出版社，2020
（世界思想宝库钥匙丛书）
ISBN 978-7-5446-6241-3

Ⅰ.①解… Ⅱ.①凯… ②娜… ③贾… Ⅲ.①物种起源-达尔文学说 Ⅳ.①Q111.2

中国版本图书馆CIP数据核字（2020）第033379号

This Chinese-English bilingual edition of *An Analysis of Charles Darwin's* On the Origin of Species is published by arrangement with Macat International Limited.
Licensed for sale throughout the world.

本书汉英双语版由Macat国际有限公司授权上海外语教育出版社有限公司出版。供在全世界范围内发行、销售。

图字：09 – 2018 – 549

出版发行：**上海外语教育出版社**
　　　　　　（上海外国语大学内）　邮编：200083
电　　话：021-65425300（总机）
电子邮箱：bookinfo@sflep.com.cn
网　　址：http://www.sflep.com
责任编辑：唐小春

印　　刷：上海信老印刷厂
开　　本：890×1240　1/32　印张 6.75　字数 138千字
版　　次：2020 年 8 月第 1 版　2020 年 8 月第 1 次印刷
印　　数：2 100 册

书　　号：ISBN 978-7-5446-6241-3
定　　价：30.00 元
　　　　本版图书如有印装质量问题，可向本社调换
　　　　质量服务热线：4008-213-263　电子邮箱：editorial@sflep.com